Mandy

BACKYARD DAIRY BOOK

the backyard dairy book
andrew singer
and
len street

Copyright 1978: Andrew Singer

Originally published in 1972 by Whole Earth Tools,
Andrew Singer's own 'backyard' publishing venture.

This third greatly extended edition is published by:
PRISM PRESS,
Stable Court,
Chalmington,
Dorchester, Dorset DT2 0HB

and printed by:

UNWIN BROTHERS LIMITED,
The Gresham Press,
Old Woking,
Surrey.

SBN 0 904727 05 X Cloth edition
SBN 0 904727 06 8 Paperback edition

ANDREW SINGER, until recently, operated with his wife
Glo a family publishing venture known as 'Whole Earth
Tools'. They did their own design, typesetting, printing,
binding, selling and distribution as a way of proving that
even 'mass communications' could be decentralised, even
down to the smallest of units. The first edition of this
book was originally published by 'Whole Earth Tools'.
Andrew is also author of The Backyard Poultry Book,
published by Prism Press.

acknowledgements

The author would like to thank the following for their help
in the production of this book:
Mr. and Mrs. B. Franche of Martock, Somerset —
champion goat keepers.
Robert and Brenda Vale of Witcham, Cambs. — ex-goat
keepers with a house cow.
Chris Bennett and Jenny Burgoyne of Soham, Cambs. —
who recently obtained a house cow.
Colin and Maggie Spooner of Chalmington, Dorset, (Colin
is one half of Prism Press) — keep goats.
Mr. R. Wilson of Fairford, Gloucestershire — keeps Dexter
cows.
Katie Thear of Practical Self-Sufficiency Magazine.

The following readers of the first edition provided useful
information which has been included in this edition:
Jim Platts of Willingham, Cambs.
Richard and Joan Collins of Westbury, Somerset.
H. Christian of Woodseaves, Staffordshire.
Many other readers of the first two editions wrote in with
useful criticisms and comments, for which many thanks.
We are hoping to receive more on this edition. Write to us
please, care of Prism Press, and it will be forwarded. If you
want a reply, please make sure that you enclose a stamped
addressed envelope.

about this book

We have written this book, without apology, as propaganda. Its aim is to provide enough basic information to encourage you, the reader, wherever you are, to start your own dairy production. More and more people are getting interested in reducing their reliance on centralised factory food production by growing their own vegetables in a garden or allotment; planting fruit trees and bushes; keeping a few chickens or having a go at beekeeping. An important element in any move towards increased self-reliance is home dairy production — so read on.

This book claims only to be an introduction. It tells you why it is worth starting dairy production; what you will need; what will be involved and how you can expect to benefit. Once it has done its job — of persuading you to take the first step — its function is finished. It then refers you to more comprehensive works for you to use as you get deeper into the subjects concerned.

We hope that you think this book is worth buying, both now and more especially after you have finished reading it. But even more, we hope that it achieves its stated purpose — it spurs you to action.

So read on .....

Andrew Singer.

contents

ABOUT THIS BOOK

1. **Why Backyard Dairying?**   1
An enthusiasm for self-sufficiency — enjoyment and satisfaction — superior quality foods.
2. **The Basic Economics of Home Dairy Production.**   5
No half-way house to increased self-reliance — what you can
save — three typical examples.
3. **A Little Technical Background**   8
How milk is produced — the lactation — letting down and
stripping out — the eight-minute oxytocin effect.
4. **What Animal to Start With**   12
The choice varies; cows, goats or sheep — points listed in
favour of each — our advice, dependant on your circumstances.
5. **Systems of Management**   19
Three systems as examples — extensive grazing — fenced —
zero grazing — choosing your system to suit your resources.
6. **Starting with a Goat: What Goat to Buy**   24
The eight breeds: their pros and cons — buying from good
milking lines — using the local Goat Society.
7. **Starting with a Goat: What else you will Need**   29
Confining the goat: the snags of tethering and hedges —
the advantages of electric fencing and how to introduce

the goat to it. Housing the goat: basic needs — ideas for construction — the straw bale shelter. Other equipment you will need: milking stand — milking bucket — optional extras — automated milking.

8. Starting with a Goat: Feeding Her.   37
Basic principles — theoretical systems of balancing the diet — the simplified backyard approach based on four guidelines — sample diet — other needs: vitamins, minerals and water — poisonous plants to avoid.

9. Milking and looking after her   44
The technique of hand milking — the routine of milking: cleaning the udder, using the strip-cup, stripping out — whether or not to feed at milking — once or twice a day? Grooming and hoof-paring. One birth per year.

10. Starting with a Cow   53
What cow to buy — help from the vet — Ministry regulations — breeds to choose from — the unbeatable Jersey. Systems of management — confining — housing — other equipment. Feeding — the basic principles — the old way — grass management — hay-making. Sharing the load.

11. Animal Problems   64
What to look out for — reference works — worm treatments — garlic as a general preventative — mastitits.

12. Dairy Production at Home   66
The traditional milk products — what they contain — how they are made: milk, cream, butter and cheese.

13. Milk and Cream   73
Straining milk and cleaning utensils — keeping milk — cream production at home — cream separators — the settling pan and syphon methods — clotted cream.

14. Buttermaking at home   78
Using different types of cream — two approaches: a quick method plus a longer one for making butter to keep.

15. Cheese production at home   86
Introducing the complexities — three ways of producing curd — soft cheeses — cottage cheese. The ten stages of making hard cheese — copying well-known cheeses — vegetarian rennet.

16. Yogurt production at home   98
Beneficial properties — Acidophilus yogurt — making
yogurt the Greek way — copying supermarket yogurt —
other methods.
17. Various other Dairy Products   103
Sweet puddings — ice cream — cooking fat — whey cheeses —
khoa — kefir.

1 why backyard dairying?

What we are proposing is that you embark on a hobby that will involve you in a fairly considerable initial outlay of money; and also in a considerable amount of work, much of it heavy and dirty and much of it outside in all weathers. Not only that, but this hobby will tie you to a regular twice-daily routine which cannot be broken and is bound to restrict you in terms of holidays, trips away and long day-trips.

So why do it? There have to be some pretty good reasons for loading yourself up with those kind of disadvantages. Here are some for starters.

An Enthusiasm for Self-Sufficiency

Since we wrote the first edition of this book back in 1972, enthusiasm for the concept of self-sufficiency has taken off. The sunday colour supplements have produced glossy guides on it; the BBC has done a comedy series on it and John Seymour (the grand old man of self-sufficiency) has written a splendid, luxurious coffee-table book on it. I suppose the glare of the media will inevitably pass, but the effect of this glare has been that the concept has become accepted and that many more people are trying to achieve some measure of increased self-reliance.

Self-sufficiency in vegetables and fruit is the first aim for most, with poultry the next step. But after that, the milk bill stands out a mile as the obvious next target. If you are one of these enthusiasts for self-sufficiency, well then, this in itself will go a long way towards offsetting those disadvantages we listed at the start in making your efforts seem worthwhile.

Enjoyment and Satisfaction

For many people, this is the prime motivation in attempting home food production There is certainly a great deal of pleasure and satisfaction to be had from keeping a house cow or house goat. When properly reared, either type of animal becomes an affectionate family pet — a member of the family. Then there is the enormous satisfaction of cheese and butter making. Both demand a high degree of personal judgement and care, but in both cases the home producer can expect to produce products greatly superior to those from the factory. Cheese making in particular, offers so many variations in technique, and many different flavours and consistencies that it can become an absorbing hobby in itself. Like wine making, it has the particular fascination of controlling and utilising a natural live process.

Superior Quality Foods

The milk you are buying right now is probably around three days old; and it has been pasteurised (heated to destroy bacteria). It may also have been left out in the sun and thus had some of its vitamins destroyed. It is probably produced by cows in large commercial herds, pushed by careful feeding of pre-mixed factory foods to their limits of milk production. Treated this way, cows are far more likely to suffer from mastitis and other udder troubles. They have to be injected in the udder with penicillin or other antibiotics. In a recent study of 40,000 samples of milk, 11% contained penicillin and 1% other antibiotics. This creates a danger of immunity to these drugs in humans and also of possible direct harmful effects on the body. One antibiotic used to treat mastitis, chloramphenical, can fatally disorganise the marrow of the bone. Surely, fresh milk from your own animal *must* be better for you than what you are getting now.

In addition, there may still be danger from the chlorinated hydrocarbon group of pesticides (DDT, Aldrin, Dieldrin etc.) which build up to high concentrations in the fat of animals grazing on treated land and can result in severe contamination of the fat in milk.

Dairy farming, like all branches of farming, has been concentrated into large units by agribusiness. The US now boasts huge 'cowtels' which have brought their own pollution problems. One of these super-farms can produce as much shit as a fair-sized town. And as you can guess, it is not 'economic' to actually use this magnificent natural fertilizer so they have been known to channel it for direct untreated dumping into nearby rivers, thus throwing out the ecological balance of large river systems.

Under the UK Public Health (Food Preservatives) Regulations, 1925-1940, the addition of preservatives is officially prohibited in all dairy products. However, there have been investigations in the USA and elsewhere of hydrogen peroxide being used to 'improve' the quality of milk for cheese making. That's the stuff that you can use to dye your hair blond — we wonder what it does to your stomach.

Apart from the pure health aspects, you may be spurred on faster to home dairy production by a little knowledge of the kind of places that now do the work previously done by pretty milkmaids in cool dairies. Butter is now widely made by a continuous process in large factories, which extrudes it out to be cut and packaged. Some cheese

3

production, notably Swiss and Cheddar, has 'progressed' to the continuous production-line stage and you can be sure that they are working on those that can't yet. But none go quite as far as processed cheese, defined as:

> The clean, sound, pasteurised product made by cominuting and blending, with the aid of heat and water, with or without the addition of salt, one or more lots of cheese into a *Homogeneous Plastic Mass.*

Processed cheese (made in slices by organisations such as Kraft) is treated with 10-20% of a special starter which inhibits certain bacteria naturally present in milk but which cause 'gas-blowing' and reduce the keeping qualities of neatly-packaged family sized cheese portions. The antibiotic used, nisin, is said not to have any harmful effects on the flora of the human stomach. Kraft slices are made by rolling presses and the extruded sheet formed is then sliced into strips, which are fed to the cutting and packing machines.

You may also be interested to know that experiments are under way on irradiating cheese with powerful X-, gamma - and other rays to prevent mould growth under plastic wrappers. So, before the dairy technologists get too carried away on their search for the perfectly uniform, infinite shelf-life dairy product, perhaps the time has come to point a rude gesture in their direction and become independent of the whole unpleasant scene.

There is no doubt that you can, if you wish, produce yourself dairy products which are superior in quality and freshness to anything you can buy. This, for some peoole, is the prime motivation in taking on the workload and responsibilities of backyard dairying.

4

2 the basic economics of home dairy production

It's all very well to be enthusiastic about the concept of self-sufficiency, but how much is all this effort going to save? Let's be clear about one thing from the start: you won't be getting dairy products *absolutely* free of charge however hard you try.

If your basic aim is to become more self-sufficient in dairy products, then we must emphasise that there is no halfway house. It's not like bread, for instance. You can save considerably on your bread bill by buying wheat in bulk and grinding and baking at home — without getting involved in growing the wheat. The same is not true of dairy products.

To make one pound of butter from bought milk, you would need the cream from ten pints or more. So the milk would cost over double what you would save in bought butter. Admittedly you would have skimmed milk as well, but this can hardly be considered the basis of increased self-sufficiency. If you started with bought cream, you would do even worse.

The picture is not so clear-cut with cheese. You need about eight pints of milk to make one pound of hard cheese, so that at present UK prices, it could just be worthwhile making your own cheese from bought retail milk. Some of you may feel like attempting to find a source of milk sold cheap because it is slightly off and so get into cheese making that way, but the point we want to make is that, generally, home dairy production only really becomes worthwhile when you have your own milking animal.

What Can You Save?

It's very difficult to do any sort of formal financial analysis of what can be saved. So much depends for a start on how much of the animal's feed has to be bought in and so much also depends on what value you put on your labour or on the extra milk you get which you would'nt have bought if you had been paying retail prices.

Let's first take the simplest case. Suppose you are a family of four with no spare grazing land but enough garden room for a goat-house plus a 10ft x 20ft concrete run. One goat kept this way wholly on bought-in feeds should provide you with your normal fresh milk needs plus a little extra in the summer for making soft cheese. You will not have enough to warrant butter-making or for hard cheeses. Of course, manure is an added bonus.

Without going into details, which will vary anyway from case to case, our analysis of costs from various sources suggests that the fresh milk you produce will cost you *roughly* half the doorstep price, with the soft cheese reckoned on as an extra small bonus now and then.

Now suppose you thought of keeping a cow in the same way. Making butter from your extra quantities of half-price milk would only be economic if you went for a Jersey or Guernsey. The milk they produce is so high in cream content that some people find it too rich to drink fresh unless it's skimmed first. So the cream and butter you get from skimming your half-price milk can be considered a free bonus. Hard cheese from half-price milk is theoretically economic, but only if you are more consistently successful at it than most backyarders I know and if you enjoy it so much that you count your labour as free. (We'll have more to say on this in Chapter 15)

The other case we will consider is when you have an acre of

land available and aim to grow all that your cow needs yourself. It can be done. The scheme consists of dividing the acre into three lots. One you use for grazing; the second for hay and the third for growing feed crops — ¼ acre oats and peas, undersown with trefoil for late summer grazing plus 1/12th acre kale and mangolds, followed by turnips. This is pretty intensive and sophisticated backyarding, but it is possible. On this, you should get 16 pints of milk per day, averaged out over the full year.

You could reckon that the sale of the annual calf should just about cover cash costs such as vet's bills, fencing, mineral licks etc., so that this system should give you 16 pints of milk daily as near as possible free of charge.

There are all sorts of intermediate cases between the above extremes, but these examples should put into perspective what savings are actually possible. Work out your weekly or monthly milk bill and work it out for yourself.

THE SHOP OF HOLLAND PARK DAIRY.

3 a little technical background

How is Milk Produced?

Milk is produced by all mammals as food for their young
during the early part of their lives when their digestive systems
are not able to take other foods. It is manufactured in the
mammary gland of the mother animal.

The simplest mammary gland known is that of the Aust-
ralian duckbill platypus, where droplets of milk ooze out of a
small hole onto the animal's hair, where they are licked off by
the baby platypus. The udders of goats and cows are far more
developed, with milk tracts leading down from the main producing tissues in the gland, contained in the milk sac, to a
reservoir just inside the teat. The teat is a sort of automatic
portion dispenser, like those used in bars for spirits, making
sure that the infant only gets down as much milk as he can
swallow comfortably. The udder of the cow is divided by thin
membranes into four quarters, each with its own teat and each
needing to be milked separately. The udder of the goat is
divided into two sections.

The process of milk production in the udder is not yet fully
understood, but it appears to be a combination of filtration of
certain constituents from the blood-stream, cell degeneration
and most important of all, synthesis of other constituents by
cell metabolism.

The Lactation

Initial production of milk in the udder is induced just before the mother is about to give birth — ready to give the baby animal its vital first meal. Actually this first milk is rather special. It is called 'beestings' (or 'beistyn'), is golden yellow and as thick as double cream.

It is designed to clear the inside tracts of the newborn animal. In England there is an old law that beestings must not be sold, but often a cow or goat will produce more than the infant animal needs. Even diluted one to four with ordinary milk, it will set like an egg custard. The taste of beestings is another added bonus for the home dairy producer.

Production of milk continues regularly in the mother animal for a period known as the lactation period. This varies considerably in length, but the point is that if you want your animal to continue giving you milk, you must have it mated regularly to make sure that it stays 'in milk' for a good proportion of its life. Although both cows and goats have been known to continue a single lactation for three years and longer, most people reckon it is safer to get the animal pregnant regularly every year.

9

An animal will only stay in milk so long as all the milk she produces is being used. This is nature's regulating system. In the normal course of events, the mother would thus only continue giving milk until its young start eating other foods. If you milk the animal dry twice every day, she will continue to produce a good quantity for longer, but you must keep milking dry — known as stripping.

A lactation really lasts as long as you reckon you're still getting enough milk to make milking worthwhile — or until you decide to give her a rest before the next birth (known as 'drying her off'). A cow calving every year should ideally have two months rest before calving. Goats seem to dry up before you decide to rest them.

'Letting Down' and 'Stripping Out'

The mother can only really be milked if the udder is 'let down'. Some outside stimulus, such as the cry of its young or the arrival of its owner, induces the production of a hormone, oxytocin, in the pituitary. It takes about one minute in a cow for the oxytocin to pass through the blood stream from the pituitary (up near the ears) to the udder, where it is believed to cause letting-down by inducing a muscular squeezing of the milk ducts, causing all milk produced up to that time to flow down to the cistern of the lower udder. Actually, it is possible to feel this filling up of the cistern happening at the start of milking.

The effect of the oxytocin wears off after about eight minutes, after which any milk produced has to wait till next milking. And then it becomes important to empty out all the milk in the cistern and teats (called 'stripping out'). They say it is also important you don't stop milking during those first eight minutes, otherwise she may only partially let down.

There's another hormone, the well-known one called adrenalin, which can complicate matters. If something causes the cow or goat's adrenalin to flow during the initial eight minutes, that reduces the flow of oxytocin-laden blood to the udder and again you get only a partial letting-down. Which is why people make all this fuss about talking to cows quietly during milking and not allowing children, dogs and the like to

frighten or excite the animal.

   Dairy farmers are neurotic about employing only competent cowmen. That's because a poor cowman, if he doesn't bother to strip the cows out regularly twice a day can dry up the production of the whole herd within weeks and ruin the farmer. The point is that if milk is left in the udder, the cow will produce less for next milking. You needn't be quite so neurotic about your own milking, but we must emphasise the technical importance of stripping out.

4 what animal to start with

By definition, all mammals produce milk, but only the larger domesticated ones are suitable as milking animals. In various parts of the world, a wide variety of animals are used — cows, goats, sheep, horses, reindeer and water-buffalo.

Climatically, Britain is not too suitable for reindeer or water-buffalo. Unless you want to go deeply into self-sufficiency and double up your milking mare as a low-cost form of transport, we also suggest you forget the horse.

That leaves us with cows, goats and sheep. Well, most of us know about cows — they produce the milk we're most used to — and in pretty large quantities. A house cow should yield around 500 gallons per year (that's 4,000 pints!).

Goats are less common, but you do occasionally see them around the countryside. The milk is less obviously creamy, and there is less of it — a 250 gallon a year milker is a good animal. There are lots of popular prejudices against goats, but most are unfounded (details later).

But who ever heard of milking sheep? Well, the famous Roquefort cheese is made of sheep's milk, as was Edam originally and although very few milk sheep are kept in this country, the idea is catching on amongst self-sufficiency enthusiasts and for good reasons. A milk-sheep yields about 200 gallons per year, but gives you meat and fleece too.

Perhaps the best way we can help to answer the question 'what animal' is by considering in turn the points in favour of each animal:

In Favour of Cows

■   Cows' milk is the milk we all know and trust. Goats' milk is virtually indistinguishable cold but when heated we find it can give off a 'goaty' smell. Sheep's milk could have the same problem, because this smell apparently is caused by short-chain fatty acids (whatever they are), which occur in sheep's milk as well, and apparently give Roquefort its characteristic 'peppery' taste compared with say, Danish Blue or Stilton.

■   Cows' milk cream settles easily and can be obtained without the use of a mechanical separator. The butter is just as you would expect it to be. Goat's milk butter can be greasy.

■   The labour costs per pint of milk are much lower than for goats or sheep. This, of course, is the reason why virtually all commerical milk production is from cows.

■   A well-reared house cow is gentle, affectionate and docile. In comparison, most goats are boisterous and capricious.

■   A cow grazes happily with little trouble to its owner and can be left out day and night, summer and winter (except in hard northern winters). A goat should have shelter at night and even during the day in case it rains.

■   A cow is easier to fence in than either goats or sheep.

■   Most breeds of cow will not destroy young trees and bushes as will goats.

■   Surplus bull calves can be sold or raised to produce veal and beef. Goat's meat is not as valuable.

■   Artificial insemination is universally available for cows, but not for goats. Goats have to be transported to the billy for mating.

In Favour of Goats

■   Goats' milk is better than cows' milk for making yogurt and hard cheese. As the fat globules don't rise, there is less chance of the cream escaping from the curd.

■   Goats' milk is said to be easier to digest. People suffering from eczema, asthma and psoriasis are often helped by changing from cow to goat milk.

■   Goats' milk will freeze without separating. (The books say you can't freeze cows' milk. You can, we've done it. But it separates and has to be whisked up after thawing to recombine the fat globules.)

■   You can sell goats' milk without a licence. In order to sell cows' milk you have to invest a lot of money in officially-recognised dairy equipment and facilities.

■   Goats are delightful animals to keep, they are full of individuality and character.

■   With a small trailer, your house goat can be taken along on family trips. A cow requires someone at home every day to milk it.

■   The milk production of a goat is about right for the milk needs of an average family. It needs less space than a cow; costs less to buy; requires less equipment and less time to look after. Even if you want more milk — for cheese etc. — keeping two goats has the advantage over one cow that you can stagger the matings and never be without fresh milk, even when one animal has to be dried off.

■   A goat will eat almost anything. It will survive on land where cows and sheep would starve.

- A goat is a more efficient converter of foodstuffs into milk. For every ten stone bag of dairy cake fed, the average goat produces two gallons more milk than a cow.
- Goats are far less susceptible to TB and brucellosis, the two major worrying diseases of cows.
- Goat kids produce meat very fast. If killed at 10-12 weeks, before they are weaned, the taste is indistinguishable from lamb, so it is said.

In Favour of Sheep

- In eating habits, the sheep seems nearer the cow than the goat. It grazes very well and has been traditionally used as an improver of grassland.
- Like a cow too, it needs no mollycoddling. It can be left out day and night in all weathers.
- In other aspects the milking sheep more closely resembles the goat. The milk is similar; the animal is small and manageable and TB and brucellosis are not a worry.
- The big pluspoint of the sheep is that it can be used as a multi-purpose self-sufficiency provider — of milk, of highly acceptable meat and of wool too — and all this on grass and little or nothing else.

That's rather a lot of points to digest. Let us sum up our advice on what animal to start with. It all depends on your circumstances. Firstly, if you or one of your family suffer from eczema, asthma or psoriasis, try a diet excluding all cow-based products (dairy and meat). If it helps, then buy a goat.

Secondly, if you have less than about an acre of decent

START

Ecxema
Asthma
Psoriasis? ──► YES

│
▼
NO
│
▼
You have less than 1 acre of grazing? ──────────────────► YES
│
▼
NO
│
▼
Is the land too poor in quality to support a cow? ──────► YES
│
▼
NO
│
▼
COW

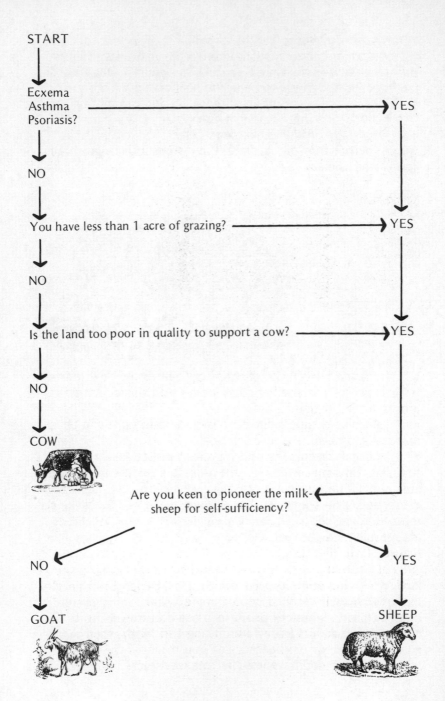

Are you keen to pioneer the milk-
sheep for self-sufficiency?

NO ◄   ► YES
│   │
▼   ▼
GOAT   SHEEP

land, you are going to have to bring large quantities of food, either gathered or bought, to a cow. The economics of home dairy production are then less attractive than if you were to use the land for a goat or a ewe.

Thirdly, if your land is of very poor quality, it will probably not support a cow, but will support a goat or sheep.

Fourthly, if your enthusiasm for self-sufficiency attracts you to join the very small band of milk-sheep keepers and so add to our presently meagre knowledge of the practicalities of this animal for backyarders, then go ahead and good luck (but please write and tell us how it works out).

But, if goats' milk doesn't solve a health problem; if you have an acre or more of good grassland available and if you are not a milk-sheep pioneer, then we recommend the good old cow as the most suitable backyard milking animal. The main reasons are, with that much good grass available, the cow makes such good use of it. Its milk is 100% acceptable and the cream is easy to obtain.

Still, a lot of those reading this will not have an acre of good grass available. It *is* possible to keep a cow on much less land by buying in all its feed and confining it to just enough space for exercise (called 'zero grazing'). But in our view this is no way for a beginner to get into backyarding. In spite of its docility, a cow is a pretty daunting animal to most of us — if only because of its sheer size. If you think of the volumes of feed and bedding required and the volume of manure to be disposed of, then you will realise what we mean.

So for those of you with less than this acre of good grazing rather than start by zero grazing a cow, we recommend starting with one or even two goats, either using the zero grazing system or others, depending on your land and preferences (see Chapter 5 for details of systems). You can then either stay with goats if you find you get hooked by the enthusiasm which affects goat-keepers, or you can graduate later to zero grazing a cow.

You can learn to milk more easily on a goat and you will make less expensive mistakes. Goats also have one unfair advantage over cows for any beginner. There are many more 'amateur' goat-keepers around than cow-keepers. They are enthusiastic, they run lots of local clubs where you can learn a great deal and they are generally very helpful to beginners.

For these reasons, after dealing generally in the next chapter with systems of management, we then go on to deal with goats first and after that go on to cows. If you want to try milk-sheep, a lot of what follows will apply, but you really ought to also read up more general sheep-keeping know-how in The Backyard Sheep Farming by Ann Williams — another title in this series.

So far as we know, the only UK supplier of a breed of sheep suitable is Mr. Lawrence Alderson of Countrywide Livestock, Eastrip House, Colerne, Chippenham, Wilts. He calls his breed the 'British Milksheep', which is a slight modification of the Dutch 'Fries Melkschaap' breed he originally imported. If you live in Holland or Northern Germany, then the Fries is the one to go for and should be avilable with a little searching. In France, the milk breed is called the Lacaune but it and the Sardinian milk-sheep give lower yields.

18

5 systems of management

Backyard milk production is essentially the conversion of what resources of land the backyarder has available into milk. The way you do this depends very much on the quality and quantity of land you have available. In other words, this largely determines the 'system of management' you use to make that conversion reasonably economic.

There are hundreds of different systems, but we can illustrate the basic choices with these three examples:

System A: Extensive Unfenced Grazing

If you are surrounded by open scrubland or moorland where a roaming goat or sheep can do little damage, you can let it do just that, providing that there is somewhere for it to shelter when it rains and at night. Your animal will gather a lot of its own food, but reckon on providing concentrates all the year round and bulk food in the winter. (For an explanation of 'bulk foods' and 'concentrates' see Chapter 8 on feeding.)

If you need land cleared, the goat is an excellent first-time clearer of brambles, briars, gorse and young trees. That's why it has been blamed for creating many of the world's deserts.

Goats and sheep kept extensively may choose to sleep out, or you may prefer to tempt them home every night with some tasty morsel or other, and thus be sure of getting your milk regularly night and morning without a hike.

System B: Fenced Grassland Grazing

This is suitable for all three types of milking animal and is the one most commonly used. All three will live happily on grass and little else when it's growing adequately, with the possible exception of some upland breeds of goat. If you intend to feed a goat mainly grass and grass hay, then go for a white goat (Saanen or British Saanen breed).

The efficiency with which you get milk out of your grassland resources can be improved by what is called 'folded' grazing — meaning dividing the available land into two or more fenced-off plots, and switching the animal from plot to plot to give the grass time to recover from a period of heavy grazing. With fencing now made cheap and simple with the use of electricity, any backyarder on grassland can practice this system.

The amount of land you need per animal depends on a number of factors: the initial quality of the grassland; how often you re-sow it; what rest you give it; whether it is expected to yield just summer grass or winter hay as well or even winter roots and other foods too. As a rule of thumb, you can start considering some sort of limited grazing system for a goat on as little as 1/3 acre and for a cow on not less than 1/2 acre — but preferably quite a lot more.

System C: Zero Grazing

A new system, labour saving and excellent if you are out all day, and one that produces big milk yields — but you should not keep a single animal this way unless it is in sight of company a lot of the time. The idea is to provide housing with feed-racks and a straw-covered sleeping area together with an outside concrete run attached. All food is brought in so no energy is wasted in gathering and very little in keeping warm. Much more is therefore converted into milk. Surprisingly, goats thrive under this system and produce splendid yields. It does not sound very labour saving but proponents claim that it is easier to mechanically cut and gather a row of grass every day than it is to deal with all the snags of fencing goats, in particular broken fences, ruined vegetable gardens etc.

The system is also applicable to cows. Many farmers keep their cows inside all winter, but we know of no-one practicing it on a backyard scale. You would need plenty of room to store feeds and bedding and somewhere to use or store the quantities of manure produced — so it is hardly suitable for the suburban semi, but could be of interest for someone with big sheds and concrete yards available and prepared to do a lot of heavy carrying work.

21

What System For You?

Between these three examples there are many gradations of system. Before you decide, answer these questions for yourself:

   — What resources do you have available — of land, buildings suitable for animal housing, concrete yards etc?

   — How can you make best use of these resources without falling into the trap that ensnares so many backyarders — over-stocking your grazing land?

Over-stocking results in the quality of the grass steadily deteriorating and also in build-up of disease organisms in the land with not enough resting periods for them to be killed off. Over-stocked land is a sad sight to behold, and its harassed owner even sadder.

Get to know what stocking rates land in your parts will sustain — ask around. Find out whether the land gets too wet in winter to take animals. If in doubt, don't plunge in too fast to giving over a lot of the land for hay or winter root production. Gradually increase hay production as you find you can afford the land.

There is no hard and fast rule how long to rest land to avoid disease build-up. If you go back very quickly, the animal will tell you by refusing to graze. If the land is in good heart with plenty of earthworms at work, this will speed up the decay

of the droppings, as will harrowing or raking them in. The longer you can afford to wait, the better. This is the argument behind the modern commercial dairy practice of dividing available land into 20 or more paddocks and moving the herd every single day to a new small paddock, stocked at about 50 cows to the acre. This can also be done on the backyarding scale by moving a simple square four post electric fence mini-paddock daily. It's cheaper than permanent paddock fencing and quicker than moving lots of big heavy corner posts every week. If you try this, reckon on a paddock 7 yd x 7 yd for a cow to start with and then adjust if it is wrong for your land. Don't use mini-paddocking for a goat, you'll never contain her!

Another thing to consider when deciding on your system is whether the land you happen to have is best turned into milk via grass or via some other feed-crop. For instance, in the west of Ireland, there can be little doubt that grass is the right use of land. But what of East Anglia, where there is less than half as much rain and often by July the grass is parched and useless? That land, or a part of it, might be better used growing field beans, kale, mangolds or various other feed-crops which better tolerate dry sunny summers. There are limitations to this approach, of course. Cows and sheep are by natture grassland animals and although adaptable, too much of some other feed can cause problems. Not too much of any one grass replacer is the answer.

6 starting with a goat: what goat to buy

Apart from choosing between the various different breeds, the first choice is whether to go for a non-pedigree 'scrub' goat through the small-ad columns of your local newspaper, or to invest in a pedigree from a reputable breeder. A 'scrub' kid or goat may cost as little as five pounds and the advantage is that if you make awful beginner's mistakes and she dies or goes out of milk, then you have not lost much. The disadvantage is that you get attached to an animal that will more than likely turn out to give an indifferent yield and you will end up doing as much work for two pints a day as you would to get eight pints from an animal in which you had invested and risked a bit more money.

As to breeds of goat, there are seven recognised by the British Goat Society and one other worthy of mention:

THE SAANEN — pure white, suitable for good-quality grazing land and giving milk around 4% butterfat content — one of the quieter breeds.

THE BRITISH SAANEN — pretty similar, but bigger and hence capable of even better yields.

THE TOGGENBURG — brown with white markings, smaller with consequently lower feed needs and yields, but butterfat content of milk is low. Suited to rougher grazing.

THE BRITISH TOGGENBURG — similar but bigger, with bigger feed intake, yield and butterfat. Full of character and energy, if you can cope with it.

THE BRITISH ALPINE — also large, black with white markings with good yield and 4% butterfat, but best characterised as bloody-minded, particularly when you try to fence it.

THE ANGLO-NUBIAN — various colours but easily recognised by its roman nose and floppy ears. A big goat giving rich 5% butterfat milk on either rough or quality grazing, but expect plenty of bleating.

THE BRITISH — this term covers goats of mixed parentage, but from pedigree lines.

THE GOLDEN GUERNSEY — this is not fully recognised by the British Goat Society. It is a sort of mini-goat and very suitable for backyarding. It stands about 25 inches to the shoulder and gives four to five pints a day, eating correspondingly less. It is very affectionate and becomes attached to its owner so only buy when it is very young. It is cheaper to buy than the other breeds but it requires special care in cold winters. There are not many about, but if you would like to try one, contact the English Golden Guernsey Goat Club (address at end of book).

Any of these breeds can be suitable for the beginner, as long as you recognise their problems, for instance, the nervousness of the Anglo-Nubian or the difficulty of controlling the British Alpine. The choice is partly one of personal preference. For instance, keenness on the mini-goat (Golden Guernsey), or on rich, creamy milk (Anglo-Nubian), or on having a pet full of character (Toggenburg or British Toggenburg). If none of these factors affect you, then we would recommend the white breeds (Saanen or British Saanen) for beginners, particularly if feeding mainly grass or hay.

Having said all that about breeds, the yield you actually obtain from your individual goat will be determined only to a limited extent by its breed. So far as yield is concerned (as opposed to factors such as temperament or size), what counts most of all is the milk-yielding capabilities of the particular male and female lines from which your goat was bred (what's called the 'strain' rather than the 'breed').

Yields vary enormously, from under two pints per day to over two gallons. It is also a harsh fact that average yields have gone down, not up in recent years. This is blamed on the fact that many breeders are more concerned with looks that will win rosettes than with milk yield. All this means that the prospective buyer has to be selective.

Nevertheless, there are a number of dedicated breeders around doing their best to improve the milk performance of British goat stock. Our strong advice is that, unless you are prepared to treat it as a short-term learning exercise only, pay more and get an animal whose pedigree promises a good milk yield.

As we said before, goat-keepers are enthusiasts and helpful to beginners. So, whatever else you do, join the local goat society. Go along to meetings and ask for advice. If you have not decided on breed, then listen to what local goat-keepers recommend. Remember too that if you later wish to breed true from your goat, you will need the services of a billy not too far away.

Apart from breeders, your contact with the local society could well reveal other members with goats for sale. But here it is even more important but often more difficult to check the milk yields of both lines from which it was bred. You pay more to a breeder of course — but there again you have the reassurance that he or she has a reputation to maintain.

Judging the likely milking quality of a kid is no matter for a beginner. That is why we emphasise getting local advice through your goat club. The best indicator, if you know how to judge it, is the kid's pedigree, but clearly you want a bright, healthy looking kid with an alert eye. The best goats for milking tend to be non-stop nervous eaters. Look also at the teats for ease of milking. A goat that has such small teats that

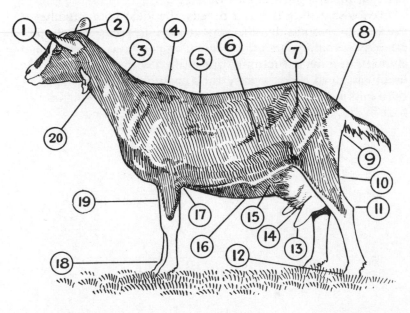

GOOD POINTS IN A MILK AND BREEDING GOAT: 1. EYE, BRIGHT AND GENTLE; 2. HEAD, SHAPELY AND INTELLIGENT; 3. NECK, LONG, NOT COARSE; 4. SHOULDERS, CLEAN AND NEAT; 5. BACK, THE LINE LONG AND LEVEL; 6. RIBS, DEEP AND WELL SPRUNG; 7. PELVIS, WIDE; 8. RUMP, SLOPING GENTLY; 9. ESCUTCHEON, WIDE AND REACHING HIGH; 10. REAR OF UDDER, WELL DEVELOPED; 11. HOCKS, WIDE APART AND STRAIGHT; 12. FEET, SOUND AND NEAT; 13. TEATS, POINTED AND DIRECTED FORWARD; 14. UDDER, SPHERICAL AND FIRMLY ATTACHED; 15. BARREL, AMPLE FOR FOOD; 16. MILK VEINS, PROMINENT; 17. BODY, DEEP, ALLOWING ROOM FOR HEART; 18. PASTERNS, FAIRLY STRAIGHT; 19. FORELEGS, STRAIGHT, SOUND, NOT TOO CLOSE; 20. THROAT, CLEAN AND FINE.

27

she can only be milked with thumb and index finger is going to slow down your milking terribly. The picture shows the points of the ideal goat. It helps, of course, if you can take along an experienced local goat-keeper to help you judge. Also, unless you want your goat for chasing scrumping boys out of your orchard, make sure that the kid is hornless or has been de-horned.

Finally, there is the question of what age at which to buy. You can buy your first goat as a weaned kid, a goatling (not yet a mother), a goat 'in kid' (ie pregnant) or a milking goat with or without kid and at any age. The ideal for a beginner, in our view, is to get a milking goat without kid. Most people can master milking technique quickly enough to cope adequately with keeping her in milk, but dealing with a kid or with an imminent birth is not so easy for a beginner. If you can't get a milking goat, the next choice would be a goatling or a goat in kid, but still a few weeks away from kidding.

7 starting with a goat: what else you'll need

Confining With A Fence Or Tether

The goat is an inquisitive animal, fond of eating all sorts of plants you would rather she did not — like prize rosebuds about to bloom, or your neighbour's field of kale — so, depending on your system, she usually needs to be confined by some sort of fence or tether. The trouble with tethering is the labour involved. You have to move her each day to a piece of ground soft enough to take the tethering stake and hard enough so that she will not pull it out. Then when it rains you have to take her in, or move a little portable shelter each day as well. In our opinion, tethering, though widely practised, is a lousy way of confining goats.

Goat-proof hedges are almost impossible to devise — due to the goat's fondness for eating most woody plants, but it is claimed that bay will keep them in, though we know of no-one who has tried it. Bay is pretty difficult to root, but once established, would make a pretty thick hedge in four to five years.

The real way is to fence. Non-electric fences have to be four foot high to contain a goat effectively, with posts every six feet. They should be constructed of sheep netting, or chain link, but not barbed wire. Now this kind of fencing has, of late, become very expensive, even if you construct it yourself. So we would only recommend it for zero-grazing exercise yards or for permanent fencing near to the goat shed. For other situations, electric fencing is the cheapest solution and probably the most secure. Its only disadvantage is the labour involved in maintaining it.

An electric fence for goats consists of three strands — one at goat-eye height; one just below goat-belly height, and one half way between the other two. Alternatively, use electrified sheep netting — more expensive but a more visible deterrent perhaps. Goats vary in their agility, but if your goat is very sprightly and can see delicious rows of cabbage in the vegetable plot just beyond her fenced field, a further strand one foot higher may be needed.

It is cheaper to make your own posts and buy just the insulators, but follow the instructions of the fencing power-unit manufacturers.

There are two types of power-unit — battery and mains. Battery units sound a better idea to most people because of the cost of running mains cable, but it is usually possible to place the unit near the mains source and use fence-wire to take electric impulses to the fence. We recommend the Rossendale mains fencer, which is based on electronic circuitry rather than mechanical relays and costs less than 30p per year to run. It is at present cheaper to buy than equivalent battery units and there is no risk of failure from unnoticed flat batteries. It was invented by a retired dairy farmer who is now a fencing-unit supplier (Rossendale Electronic Fencers, Rossendale, Lancs). Whatever you do, don't waste money on a fencer that gives a weak shock only. Goats take big shocks to persuade them not to attempt to break out.

To get your goat to respect the electric fence, you will have to wait until it is beyond the playful kid stage. Then get a neighbour or friend to hold its mouth around the fence wire for three or four good jolts. Unless it is a very stubborn goat, it will not go within a foot of the wire again — ever. Nor will it go near the person who held it — that is why it is not a job to do yourself.

The problem with electric fences is that if the bottom strand touches wet grass, it earths and wastes electricity continuously. Cows conveniently reach under the fence and munch as far as they are able to reach without getting a shock, but goats are too careful to risk that kind of unpleasantness. So you have to go round with a pair of garden shears every so often cutting back potential earthing grass. Some power-units, including the Rossendale, flash lights when earthing is happening, so if you check every day, you need only to cut back the grass when it is actually causing a short circuit.

Housing Your Goat

SHELTER OF POLES AND BOARDS (1893).

If you are using zero grazing, you will need a large goat house with room for the animal to wander round — a minimum of one hundred square feet. Otherwise, if the house is merely to be used as a rain shelter and sleeping quarters a goat can manage with something a quarter the size. But you will, of course, need somewhere to store bulk food like hay and concentrates (best stored in plastic dustbins so as to discourage rats and mice). Unless you are very hardy, you will also need somewhere under cover to do your milking. If you are having to make or buy goat housing anyway, it makes sense to allow enough room for storage and milking while you're at it.

Milking is done on a stand (see illustrations) and takes up very little room. A corner 5 ft x 4 ft would do. Two plastic dustbins likewise take up very little room, but make sure it is 100% impossible for the goat to reach them. She will easily kill herself with over-eating if she gets in. As for the hay, it is best stored in a loft above the goat quarters, since this helps to insulate against cold and makes it easy to fill hay-holders by dropping hay down through a hole in the loft floor.

A lot of the designs for goat houses illustrated in specialist goat books are meant for keeping lots and lots of goats, where a prime consideration is the avoidance of bullying and making sure all goats get their full ration of food. If you are keeping one or two goats only, you can dispense with the fancy stalls, with special bars for the goat to poke her head through to reach the feed bucket.

In designing your goat house, be aware of the goat's needs in the matter of shelter. A goat's natural method of keeping warm in the wild is not to snuggle into a deep bed of straw like a hamster, but to find a dry, draught-free shelter and keep warm by eating lots of roughage and converting it to internal energy.

Most books on animal husbandry say that housing should be draught-free but provide plenty of ventilation. How to achieve this amazing combination is rarely explained. The secret, in our view, is to provide adequate insulation and to ventilate only one wall.

Draughts are set up when air warmed by the animal is cooled by contact with cold walls or ceilings. Insulation avoids this. Ventilation should be adjustable and not blast directly onto the animal. It doesn't matter that the air around the animal is cool, but it should not be circulating fast as a draught and nor should it be stale and unmoving.

So if you construct the house of wood, it should be double walled with some insulation between. The floor is easiest if made of concrete. If you lay a damp-course below the concrete and lay a thin top layer of strong concrete directly over a two inch layer of polystyrene insulation, you can get away without straw bedding. The concrete floor is then warm enough for the goat to lie down on directly. If the floor is made to slope gently to a gully at one wall, which itself drains outside the shed into either a septic tank or a regularly-emptied garden manure tank, then the floor can be swept clean every morning with little trouble. Mucking out straw bedding is a job you do not have to do often, but it is a messy business. Even if you do use straw, though, the slope-plus-gully design helps to keep the straw dry for longer.

The housing should include provision for feeding concentrates, bulk feed and drinking water. Buckets for concentrates and water should be held so that they cannot be kicked over or trodden in. Bulk feeds are best fed in wall mounted feed racks — made out of rope netting, wood slats or wide metal wire-mesh — all of which allow the goat to pull down hay etc. in a way that comes naturally to a tree browsing animal.

When designing your feed racks, there is one detail you should allow for. Goats are fussy — they even ignore the tastiest food once it has fallen on the floor, yet having no hands to manipulate hay and the like, they inevitably drop a sizeable proportion of what they pull down to eat. The answer is to arrange for any food that drops from the rack to fall where it cannot be fouled, and where it can be collected later to be put back up in the racks. This detail will save you a small fortune in feed bills.

To finish this section, here's an idea for cheap basic, do-it-yourself housing — a very basic goat shelter made of straw bales. You construct the shelter rather like a Lego house, with the cut ends of the straw pointing to the inside of the shelter. Each added bale is tied in to its neighbours with twine. The roof bales are laid across supports of scrap wood (old floor boards, rafters etc.) and then covered with strong agricultural grade polythene sheeting or corrugated iron, itself tied down to the bale structure. Line the structure on the inside with strong wire mesh, not chicken wire, so the goat doesn't eat its house down, and leave one side completely open — preferably away from the prevailing winds.

The floor is more effective if dug away and filled first with

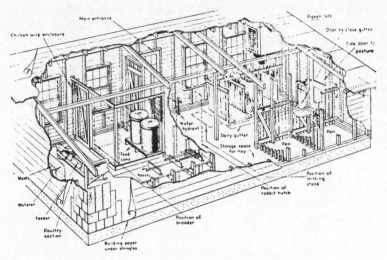

large stones, then smaller ones — this allows urine to soak away. The bales last longer if placed on a bed of bricks or stones too.

Then you have it — the ultimate in dairy self-sufficiency; a home-grown goat shelter. It will last three to four years, after which burn the bales and blowlamp the other bits to avoid carry-over of parasites to the new one.

34

Other Equipment You Will Need

Goats are rather low animals, so we recommend some sort of platform for milking. It should be at easy sitting height and just big enough for the goat to stand on comfortably, but not big enough to let her move around and make milking difficult There should be some arrangement for holding or tying her head at one end. The illustration is from the U.S. Department of Agriculture Leaflet No.538 and shows a pretty sophisticated design. The bolts at the top of the stanchions pull out to get the goat's head in. One fixed stanchion and one swingout stanchion would be a slight simplification. Also, you could do without the ramp, since goats are more than happy to negotiate a step of 18 inches. The timber is 4 inches x 4 inches for the legs, 2 inches x 4 inches for leg-braces and framing, otherwise 1 inch boards.

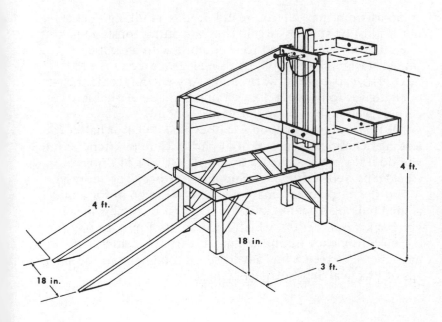

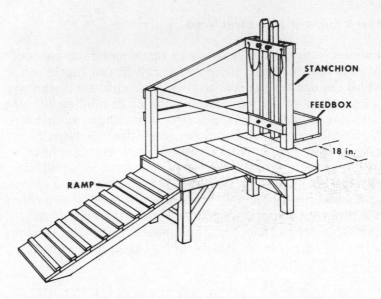

STANCHION

FEEDBOX

18 in.

RAMP

Apart from that, all you really need is a milking bucket, but it must be stainless steel. They are outrageously expensive but don't be tempted to economise with a plastic or tinned steel bucket. You'll be using it twice a day for the next few years to hold warm milk. Any encouragement that bucket gives to bacteria, like scratches and unwashable corners, will make you regret your economy.

'Optional extras' include a leather goat collar or halter for leading her about, a special hoof-paring knife (a kitchen knife will do if it's really sharp), etc. etc. These sort of things are stocked by Self-Sufficiency Supplies (see back for address).

For the modern automated goat, you can go in for a one-animal milking machine, but the cheapest of these will set you back around £400, which needs a lot of milk to pay for it. Self-Sufficiency Supplies import it from Belgium. Alternatively we've seen a few advertised second-hand in the small-ads of Practical Self-Sufficiency magazine.

36

8 starting with a goat: feeding her

Basic Principles

The goat breeds used in this country are nearly all based on breeds from the Swiss Alps, where conditions favour tough, fibrous foods with low food values. Admittedly, some of the breeds have been modified into lowland goats, more suited to richer, lusher, vegetation, (e.g. the Saanen) but most have not.

Our goats are browsers by nature, not grazers. They are used to conditions where they have to search out food over distances. They are used to having to reach up for the food.

Goats do not have thick furs, considering they come from cold mountain areas. They keep warm by eating fibre and converting it internally to heat energy.

There is very little that grows that goats won't eat. On unrestricted range grazing, they can be expected to browse for themselves a pretty balanced diet. But, if restricted, then you have to see to it that the diet is adequate, varied and balanced.

Theoretical Systems Of Balancing The Diet

There have been various approaches used to simplify animal nutrition down to a few simple factors so as to allow each farmer or goat keeper to balance a diet according to what he has available.

The well-established approach is to say that a goat (or cow, or sheep for that matter) uses food for two purposes. Firstly for the maintenance of its own body and bodily functions (called the maintenance ration) and secondly for the production of milk (called the production ration). The next simplification is to say that the maintenance ration needed is proportional to body weight and the production ration to the amount of milk being produced. The whole thing, of course, is a gross over-simplification, but the purpose is to provide a rough guide.

Now human rations are usually expressed in calories — measuring energy value and also in the weights of proteins, carbohydrates and fats which are the three main constituents in human food. Animal rations are traditionally analysed with respect firstly to the starch content, being what provides energy and secondly to the protein content — to build milk and flesh.

So, you will read in Mackenzie's 'Goat Husbandry' that the goat requires daily:

0.9 lb of 'starch equivalent' per 100 lb bodyweight plus 0.09 lb of digestible protein per 100 lb bodyweight (for maintenance)

and 3.25 lb of starch equivalent per gallon of milk plus 0.5 lb of digestible protein per gallon of milk (for production).

There are tables giving starch and protein contents of various foods so you can work out what all this means for your goat.

There are all sorts of things wrong with this approach. Firstly, what about the fibre which, as we have seen, is the major energy constituent of the goat's diet? Secondly, what about the element of variety itself, which is so much part of the goat's natural diet?

There is no ideal simplification. Recently, cow experts have gone over from starch and protein figures to a single figure measuring the energy value of foods in 'Megajoules'. And so the show goes on. But what of the poor simple beginner? How should he go about balancing his goat's diet? We suggest that you leave the scientific stuff for later and start off by following a few simple guidelines.

Backyard Principle No.1: Feed Maximum Variety

Whatever your land conditions, give the goat as much of a mix as possible. Make your grazing as interesting for her as you can. It has been proved that grassland with a large mix of plant types has much more mineral value than pure grass. On restricted grazing, a goat should if possible, have access as well to some tree and shrub leaves — particularly in winter when she needs the fibre to keep warm.

You are likely to need to supplement this with crop feeds, particularly in winter. Potatoes, carrots, kale, drum-head cabbage are all acceptable. Again, aim for variety.

You may need to buy in hay just for maintenance through the winter — you mustn't economise on that. But buying in food to give extra milk needs to be looked at carefully. Most books will assume a continual diet for your goat of bulk feed (grazing or hay) plus concentrates (grains, meal etc.). This may not be worth your while. It may be more economic sense to run your goat through the winter at somewhat below her potential milk yield rather than pay out a lot for concentrates.

Backyard Principle No.3: Vary it with the Season

All this neat 'starch and protein in — milk out' stuff is misleading. No goat will give you a steady return of milk throughout the year, however steady your rations. Nature doesn't operate like that. Through the winter, keep the diet dry and full of fibre and gently rising in quantity till you dry the goat off eight weeks before kidding. Then keep her at maximum rations till three to four weeks after kidding and gradually reduce in time with the natural reduction in milk yield that operates as the lactation proceeds.

Backyard Principle No.4: Vary Quantities as to the Look of the Goat

The most direct guide as to whether you are feeding too little or too much is the goat herself. Get to know her normal healthy look. Feel round her ribs and along her back regularly to get used to how it should feel. You'll then more easily recognise signs of over or under feeding. Recognising specific mineral deficiencies needs more skill and you need reference material beyond the scope of this introductory book. Here is

40

the place for us to recommend, of all the specialist goat reference books now on the market, David Mackenzie's 'Goat Husbandry' is the best of them all. If anything, it's almost too detailed for the backyarder, but as a work of reference, once you get into goat keeping seriously, it is unsurpassed.

Those are our principles. Ask the person you buy the goat from what its rations have been. Gradually modify these to suit the resources you have available and watch out for sudden drops in yield — you may dry her off without meaning to. Then when things have settled down, experiment, if you wish, with reducing concentrates — particularly when the vegetation is lush in early summer.

Just to give you something against which to compare the advice of the person you buy your goat from, here is a sample diet for a reasonably high-yielding goat:

1—2 lb concentrates, depending on the season and possibly dropping lower if other feeds suffice — the easiest way is to buy 'goat mixture' from a local seed merchant or from a local breeder who mixes it. If you can't get goat mixture buy ewe and lamb meal in preference to dairy nuts). If you want to go into it a bit deeper and start saving money, you can use mixtures of some protein-rich concentrate like soya meal, kibbled beans, oil-cake, cornflower residue meal, dry brewers' grains mixed in with starchy concentrates like bran, sugar-beet pulp or any of the cereals. The aim is to get at least 12% protein and a decent variety in the mix.

Either plenty of mixed herbage and shrub grazing to provide the bulk and fibre needs, or, when not enough of this is available, hay and feed roots in virtually any combination — but 5 lb of hay with 5 lb of roots could be enough where absolutely no grazing is available, depending on the goat's size and yield of course.

Other Needs: Vitamins, Minerals And Water

Vitamins don't often seem to be a problem in goat keeping, but minerals can be. The greatest need is for salt (15-20 lb of it per year) followed by calcium and phosphorus in the right balance and then a load of trace needs like magnesium (which is a seasonal problem with new spring grass), copper, cobalt and others.

The simplest backyard approach is to buy a mineral lick from a corn or farm feed merchant. This is a great big block containing the right mix of minerals for a cow — not ideal for a goat but it will do. It is hung up under cover where the goat can *just* reach it for an occasional lick when she feels the need. If it's put any nearer, she'll playfully destroy it and these licks are not cheap.

Some of the mineral problems of goats are problems of inbalance between particular minerals rather than of deficiency. For that reason, more sophisticated goat keepers favour separate supplies of the various minerals placed in little boxes in the goat house so the goats can mix their own and supposedly get it right by instinct (we're not convinced).

Finally, don't forget water. It must be clean and continually available to the goat, preferably in a form of container she cannot knock over or foul by climbing on. A bucket is a very poor receptacle. A proper trough with ball-valve is the ideal. In winter, it's as well not to let the water get too cold, so supply it if possible in the shelter or goat house.

Poisonous Plants

You should be aware that the following plants are poisonous to goats and cows:
- Laburnum
- Rhododendron
- Yew
- Beet and Mangold leaves (unless wilted first)

If you suspect poisoning, call the vet right away. Many plants affect the taste of milk, butter and cheese. If you suspect that your milk is tainted, the Ministry of Agriculture Bulletin No. 1616 might be of help to you. In it there is a chart which relates various plants to the effect they have on milk, butter and cheese.

9 milking and looking after her

Milking is fairly simple to master on most cows or goats, although there are individual animals that are very difficult milkers. The principle is to block off the teat, full of milk, from the udder by squeezing with the index finger and thumb, then to follow this with a squeezing downwards and out of the milk trapped in the teat, using the other three fingers. Unless you cut off the milk in the teat from getting back into the udder, no amount of squeezing will get much milk to flow. After practice, the action simplifies into an even squeezing downwards from index to little finger. The trouble is that this feels wrong at first. The natural tendency is to want to start squeezing with the little finger. Once you have got over this, you should be alright. If not, you may have a difficult animal and therefore you should seek advice.

The technique varies slightly from cows to goats or sheep. Most cow milkers pull down gently as they squeeze out the milk, though it is not entirely necessary unless the teat is too big for you to squeeze out without pulling. You certainly should *not* pull on a goat or sheep.

The drawings show an ideal teat. Most goat and sheep teats and some cow ones will be smaller and fiddlier to milk. The technique is the same — you just miss out the fingers that overlap the teat. One more thing — keep your fingernails short and clip the udder if it's so hairy that milking is awkward.

First relax grip and push up — then start
pulling down as you squeeze first finger
second finger-last two.

The Routine Of Milking

Having read the mechanics of letting-down in Chapter 3, it will
be apparent to you how important it is to stick to a regular
routine of milking, readily recognisable to the animal so as to
stimulate letting-down and totally free of stresses or hurry, so
as not to stimulate adrenalin flow. This is not too difficult
with the placidly disposed cow, but goats can be very twitchy
and neurotic and cause chaos at milking time very easily:
kicking over milking buckets is a favourite one.

Most people feed concentrates at milking. This seems to
have the advantage of keeping the animal occupied and en-
couraging her to come for milking without any bother. But
generally a cow or goat full of milk is glad to have the swollen
udder relieved of its load, so such an extra incentive should
not be necessary. Some dairy farmers are beginning to question
the logic of feeding at milking time. After all, they say, a cow
always stops feeding to suckle her calf and although a cow
dungs when feeding, she never dungs when suckling — so if

you don't feed at milking, your milking area stays clean and you're operating more in tune with natural conditions.

This seems to make good sense. Even more, perhaps, for goats. Without the feeding, perhaps some of their excitability would be curbed and make for calmer milkings. The problem is that changing the routine in a mature goat is a nightmare, so no-one we know has yet tried it out on a backyard scale. But if you are starting with a newly-kidded first-time milker, it could be worth a try. If you do, please let us know how you get on.

If you don't feed at milking time, it either means an additional trip out to the goat some other time or you do it before milking. If you leave it till immediately after milking, the excitability would be worse.

Before you start milking a cow or sheep, it is usually advisable to wipe the udder with a damp cloth kept expressly for the purpose and washed out daily. This stops muck on the udder falling in the milk. Goats keep their udders so clean this is not necessary.

The actual routine of milking should start with milking out the stale milk sitting in the teat itself and letting it go outside the bucket. Commercial cowmen catch it in a 'strip cup' so that it can be examined for white spots and clots, which signify mastitis. This is less vital in the less high-pressure conditions of backyarding.

By that time, the animal should have let down and you can start milking. Don't be tempted to learn by using one hand only — you will never progress. On a cow you milk two teats in turn — the two nearer you then the two the other side.

Keep milking with as little break as possible until the supply starts to ease off. Then you need to strip out. We like to alternate massaging the udder with both hands with a turn at milking, but the other way is to massage each quarter in turn with one hand while milking it out with the other. Massaging in this context means coaxing the remaining milk down from the upper to the lower udder with a downwards motion.

Beginners find it difficult to judge when to stop, because there seems always to be a drop more that could be coaxed out. Once you get to the stage where an additional massaging brings

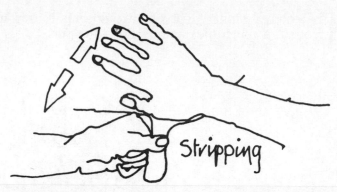

Stripping

you less than three squirts, we reckon she's stripped. But to be sure, get an experienced milker to check out that you are stripping adequately, if at all possible.

Although goats, sheep and cows produce best if milked twice a day, as near to every twelve hours as possible, this isn't essential. Milking at 9 a.m. and 6 p.m. is acceptable, but you'll get slightly less milk. You can even milk just once a day if you wish and make do with even less production. A half-way house to this approach is to use a cow partly as a milker and partly to suckle and rear calves for eating. You take the morning milk, having kept the calf out of reach of the udder overnight and let the calf take the evening milk.

Looking After Her

There is not much to grooming a cow or goat. Brushing with a dog brush will be appreciated by both animals and help to keep a sheen on their coats. Grooming helps to strengthen the relationship between you and the animal. After all, she is giving you milk as if to her young and should be even more ready to do so if she feels good towards you.

A goat's hooves need care, since the ground you keep her on is likely to be softer than her rocky natural habitat. The hooves will grow too long, rather like toe-nails, and need cutting back every now and then. If the front of the hoof starts to curve up, it certainly needs cutting. Use a sharp kitchen knife to pare the hoof back to where it leaves the ground.

Cow's hooves similarly need paring but only if they are on soft ground. A rough file or rasp will do the job.

One Birth Per Year

Although it is possible to 'run through' a lactation into a second year, it is not usually considered economic. You lose the benefit of a calf or kid to eat or sell and the milk yield drops steadily through the lactation. So how do you do your family planning?

The goat carries its young for 150 days. It is generally reckoned that April and May are the best months to arrange a kid's birth in that herbage is building up. So a successful mating in November or December would do the trick. But matings are not always successful and other things can delay visits to the billy, so it is as well to start watching out for the goat being in season in September or October. You can then take her to the billy immediately or make a note that she will come into season again three weeks later.

The cow carries its calf for 280 days and a birth around April or May is again a good idea. So, working back, a July mating would be about right. Once you get past the cow's first calving, you can keep to an annual cycle, starting to watch for her being in season 60 days after calving and mating ideally at 85 days after to make for a regular 365 day cycle.

progressing to a cow

10 starting with a cow

Rather than repeat a whole lot of things, we are concentrating in this chapter on points of difference between a cow and a goat. So, if you are starting off from scratch with a cow, begin by reading the previous chapters on goats before you come to this one.

What Cow To Buy?

We would not recommend the inexperienced to get his first cow from a cattle market. There are no cow equivalents of the goat societies listed, although you may well, as a goatkeeper, have made contact with house cow owners. The other good source of help is the local agricultural vet who you may also have met as a goatkeeper. He will know the local dairy herds, and if you are lucky, could put you in touch with someone with a suitable cow to sell. An idea worth considering is to start with a fairly elderly cow from a commercial herd. She is perhaps no longer giving enough milk for the herdsman to keep her on, but she could be ideal for you. Scout around and see what you can find. If the vet is unhelpful, try another. The vet will help you, too, in making sure that the cow or calf is TT tested, vaccinated against brucellosis and free of mastitis. Do not be put off by all this talk of diseases since they are all fairly well under control now.

The reason why they are fairly well under control is, no doubt, because of some pretty strict rules applying to cows. If you are getting a cow, it's as well that you understand at the beginning the regulations to which you will need to adhere.

Brucellosis is a disease of contagious abortion. The Ministry of Agriculture is gradually clearing the UK area by area. If you live in an 'eradication area', regular tests are compulsory and no cow can be moved in without a permit and without coming from a 'Brucellosis accredited' herd or without having a blood test done. All this is pretty sensible stuff, particularly as the disease can affect humans as well. Apply to your local Divisional Veterinary Officer at the Ministry if you need advice or further details.

Tuberculosis has been virtually eradicated by previous similar controls, but all cow owners are still obliged to ear-mark each cow on the right ear with a number,which will be allocated to you by the Divisional Veterinary Officer.

Mastitis is an infection of the udder. There are no obligatory measures in force as yet, though the Ministry can advise you on precautionary techniques.

While we are on the subject of official regulations, all cows need a permit to be moved. If you have an 'accredited herd', you can enter up your own permit book. Otherwise, apply to the Ministry's local office.

Hilariously, the control of sheep movements is not a Ministry local office job but a local authority responsibility which, in our area at least, is given to the Office of Consumer Protection! If sheep scab gets a problem, they supervise the obligatory dipping.

Breeds To Choose From

There are quite a number of regional breeds of cattle — some selected over generations for optimum meat production, some for milk and some known as dual-purpose. The Jersey cow is worldwide the most popular milk-producing breed, but in this country, three quarters of our doorstep milk comes from the Friesian which, strangely enough, originated from the same islands as the Fries Melkschaap. The Jersey, you see, shines as a producer of butterfat, so places like New Zealand use this

breed to turn their fine grassland into exportable butter. The
Friesian, on the other hand, shines as a producer of milk volume, so where milk is sold fresh by volume, it is the animal
used.

If you have good quality grassland available and intend, as
you are almost obliged to with a cow, to turn some of the milk
into cream, butter, ice cream and other goodies, then the Jersey is without doubt the animal for you. It is by far the most
popular breed of house cow. It is small enough to handle as it
is about two thirds the size of a Friesian, and gives an unbeatable 5% butterfat-content milk. It is also high in milk proteins
and milk sugar and although the yield is lower, an average of
two gallons a day, it is more than enough for most families
today, even if they make cheese regularly through the summer.

The Jersey also has unbeatable looks, is docile and becomes
attached to its owners in a very engaging way. Its only possible
snag is that pure Jersey calves do not produce saleable meat.
There is nothing wrong with it, but the fat is yellow and no-
one will buy it. Also, the Jersey calf, being a milk breed, does
not produce meat particularly fast or economically.

You can get round this by mating your Jersey to a meat
breed. The Aberdeen Angus is often used. This gives more
meat and white fat, but the little calf's Highland origins make
him very difficult to fence in. You can mate to other meat

breeds, but take advice from the local A.I. (artificial insemination) people first. Mating the little Jersey to too large a breed can cause serious calving difficulties, as you can imagine.

As for other breeds, the second choice might be a Guernsey. It is very similar but slightly bigger and it gives better yields although the milk is less creamy. However the cream content is still above the average. It is said to be less temperamental than a Jersey, although we have yet to hear of temperament trouble in a Jersey.

There is also a cow equivalent of the mini-goat breed we have already described. This is the Dexter, a mini-cow originating from the west of Ireland. On average, it stands only 39 inches to the shoulder (about level with your hip) and it is very hardy, adapting to all sorts of grazing. So, it would survive on land that no Jersey or Guernsey could cope with. The milk is about halfway between doorstep Friesian and Jersey standards and comes in adequate quantities of 1½ to 2 gallons per day. The beef is good too and comes in nice small joints.

The ideal house cow, you may be thinking. Alas, not so, though one day it might be. The Dexter is very much an enthusiast's cow, with limited commercial value, so the numbers are very low — only about 40 to 50 calves are registered each year with the breed society. This has two effects. Firstly, it costs about three times as much to buy a Dexter as a Jersey, at present. Secondly, inbreeding and breeding for smallness have taken their toll. The breed is reckoned to be temperamental and still subject to breeding problems, though the breed society claims these are being reduced. About one in six of the calves are born deformed.

We don't know of anyone who has tried crossing a Dexter with a Jersey to create an improved mini-cow for the backyard but it seems an obvious thing to try. The address of the breed society is at the back of the book and they offer an AI service.

SYSTEMS OF MANAGEMENT

The choices in chapter 6 apply equally to cows, though rough moorland grazing would only really work for a Dexter, and even then would have to be supplemented to get a good yield.

Confining

N.Y. Public Library Picture Collection
Pens and frame of archway for a shelter (1893).

Cows are less likely than goats to be attracted out of a field by tasty bushes and trees. But, if they are lonely or on heat and can hear other cows, they can be almost as energetic in their attempts to get out. Give a single house cow plenty of your company and affection and keep her well confined during heat. Ordinary farm fences and hedges should then be enough, although electric fencing allows you greater flexibility in rotating your available grazing.

Cows need less drastic training onto electric fences than goats. Put some tasty food on the other side of the fence and watch. A few jolts and she will get the message. Also, a single strand at about the height of her nose will be adequate. If there is a calf in the field, though, add another at 18 inches.

Housing For Your Cow

So long as the ground does not get impossibly soggy and mud-clogged, and so long as the winters are not too harsh, most cows can stay out day and night throughout the year. Milking can also be done outside but a rain shelter is useful. However, if you keep your cow this way, she clearly has to eat more to

INTERIOR OF IMPROVED LONDON COW-SHED.

keep warm (maintenance ration). If that is not a problem for you, even in winter, then this might be the right approach.

Otherwise you will need housing. The cow needs much the same sort of facilities as the goat — just larger. Also, the manure is a lot wetter, so if your housing will use straw bedding, then make the bottoms of the side walls out of brick or breeze blocks up to about 3 ft high, so that the soggy under-layers of bedding do not rot the walls.

Other Equipment

Needless to say, no milking stand is required (even for a Dexter), but you will need a milking stool to get yourself down to milking level. Buy or make the old-fashioned three legged milking stool. Apart from that, you will need a milking bucket, an udder cloth and a leather or chain collar or rope halter for leading the cow.

Feeding

Cows are grazers, not browsers. They are designed to eat grass and grassland weeds ad nauseum, moving slowly and steadily over the ground available. So it is quite possible to feed a cow

59

on summer grazing, winter hay and nothing else, so long as the grass is of good quality and contains a good variety of plants.

The four feeding principles of chapter 8 apply to cows:

— Feed maximum variety and although this is not as critical it helps to keep up mineral supply.
— Be economical in bought-in feed. This is even more important for a house cow, since buying concentrates to turn into winter butter and cheese is not very economical. Better to make plenty from the spring grass, freeze some for winter and just milk once a day through the winter for fresh milk only.
— Vary it with the season. The same pattern applies, though for the most part, the cow does the adjustments herself.
— Vary the quantity with the look of the cow. This matters in winter when you may decide to cut right back. If she is in calf at this time, she should not be allowed to run down badly.

Grass is basically a spring growth, with some in autumn as well. The rest of the year, it makes little or no growth. To make the best of this, the old pattern of farming was to pro-

duce a calf in spring, get milk through the summer from that grass growth, then dry off after the grass stopped growing and keep the cow just going on roughage and little else through the winter — with the calf providing winter meat for the family instead of milk.

This pattern has been totally changed in modern commercial farming, in order to produce as steady a supply as possible of fresh milk, making use of what has, up till recently, been cheap imported concentrate to do so.

Now that concentrates are no longer cheap and freezers are widely available to store milk, butter and cheeses, there is a strong argument for the backyarder to return to the old pattern.

If you manage the cow this way, you can winter-feed on hay, if you have enough grass to make it, or experiment with low-grade foods just to keep her going — like oat or barley straws, beet tops etc.

Grass Management

This is a subject deserving of a whole book (and there are some if you are interested), but we just want to give you a few beginners guidelines.

— Don't let the cow onto the spring grass until this has grown to a height of at least 3 to 4 inches, otherwise she will ruin the spring growth.

— Ideally, you should re-seed pasture at least every four years.

— Remember to go for variety.

— Remember to rest the land in turns.

— Like your garden, grassland can benefit enormously from liming. Do a simple garden kit test on it now and then, and lime accordingly.

— A test on your land's need for nutrients may also be worthwhile.

— If you can organise it that way, rotating land between grass and crops also helps.

Hay Making

Hay making is giving way in large measure to silage. Although more machinery is involved, the farmer is less of a slave to the weather. For the backyarder, however, silage is not on. The outer layer is nearly always spoilt, so that small-scale silage making is uneconomic compared to hay making.

The skill in hay making is choosing your time. You should cut at the start of a period of dry weather, so that after a couple of turnings, the hay has dried out and is ready to be stored. The time of cutting is also a compromise between two opposing considerations:

— the older that grass gets, the less nutritious it is.

— the taller the grass, the easier it is to handle as hay.

In the Swiss Alps, apparently, they cut hay at 6 inches and consequently make a very nutritious winter feed, but goodness knows how they handle it — perhaps with a lawn mower. Most backyarders either make their own hay by hand or ask a neighbouring farmer to bring in his baler. Particularly if you are turning and loading by hand, the grass has to be at least a foot long to tangle nicely and be easy to pick up with a hayrake or fork.

Road verges are potential free sources of hay. If you catch the council man when he is mowing a verge, he may be more than happy to let you turn it and take the hay away. Country

lanes are alright, but the verges on main roads are likely to be too full of lead deposits from the exhausts to be good for your cow.

A last word on feeding. Unfortunately there is no bovine equivalent to Mackenzie's *Goat Husbandry*. The book that comes nearest in our opinion is Newman Turner's *Herdsmanship*, also published by Faber and Faber. Although this book is designed for the commercial dairy herd owner, its approach is organic and there is plenty of useful information for the one cow beginner. This book is now out of print but your public library should be able to locate a copy for you. There is also quite a nice recent American book called *The Family Cow* by Dirk van Loon, but naturally enough, a lot of it is very specific to U.S. conditions. If you want to try scientific feeding, get hold of Ministry of Agriculture Technical Bulletin No. 33: *Energy Allowances and Feeding Systems for Ruminants*, which details the Megajoules approach to feeding.

Sharing the Load

We have already emphasised what a tie it is having either a cow or a goat. Having a cow, you are also faced with having to make use of pretty large quantities of milk. Unless you have got all the attributes of an officially approved dairy, you are not permitted to sell the milk. But what is wrong with owning a cow in partnership with friends? A good scheme we know of is run by two couples, one with two acres and the other in the next town. They share the milk and the feed bill but, for their work, the country people get the calf every year as well. The only drawback to this idea is that you must have some method of getting the milk to your partners cheaply at least every other day.

11 animal problems

It is beyond the ambitions of this book to give you a reference list covering symptoms and treatment of all the various diseases that goats and cows can get. If that is what you want, get a copy of Mackenzie's *Goat Husbandry* or van Loon's *Family Cow* or Norman Brown's *Dairy Farmer's Veterinary Book*.

Furthermore, illness comes so rarely to the backyarder so why should he even try to become an animal health expert? In our view, the important thing is to get to know the normal healthy look of your animal and train yourself to be sensitive to changes.

Look out, in particular, for:
— listlessness
— lack of appetite
— a sudden unexplained drop in milk yield
— lack of cudding
— discharges of any sort
— excessive salivation or coughing
— swollen joints or signs of lameness
— scouring (the technical term for diarrhoea in animals)

With one animal that you see every day, keeping a watch becomes second nature. But don't overprotect. Give a little time if the change isn't serious. Vets are expensive, but only if you call them out. A quick bit of advice over the phone is usually free.

Some people favour regular de-worming with thibenzole

tablets; others reckon that a better policy is merely to keep worms in check but not to destroy the worm population completely. The idea is that an animal with no worms at all will have no resistance to a sudden attack.

The best protection against worms is to switch fields as often as possible. But regular use of garlic or garlic tablets is the other way of keeping worms in check. Garlic seems to have almost magical properties in keeping stock healthy and is certainly pretty cheap, though some claim that its taste gets through to the milk.

Mastitis can be a problem with milking animals. Take care to avoid udder injuries, but treat them and milk gently when they do occur. The Ministry is promoting a new antibiotic routine which you carry out at drying off, which reduces risk of mastitis by 70 per cent, but few hand milkers have much trouble with mastitis anyhow.

Apart from that, the rest is mostly common sense. If changes happen that worry you, contact your vet. If you have a sudden unexplained death, be sure to get the vet or the Ministry to check for Anthrax, which is a serious and notifiable disease (that means you can be fined if you don't inform the Ministry). If you discover your cow lying on her side unable to get up, roll her over from time to time while you wait for the vet to come.

12 dairy production at home

Milk is a pretty good food as it is. In fact, there are some primitive people who live on virtually nothing else. So why bother converting it into other things? The traditional reason must have been to produce from it foods that could be kept for the winter, but nowadays this reason hardly applies since adequate supplies of milk are commercially available all the year round. No, butter and cheese are still made today because they represent concentrated sources of fats and proteins which it is economic to transport over longer distances than milk and because they have their own consistencies and flavours prized independently of milk.

The backyard dairyman can, if he wishes, obtain excellent year-round food from milk alone (assuming he has plenty of winter feed for his animal, or a big freezer) but butter and cheese will further enrich his life. Not only that, but instead of throwing the inevitable surplus summer milk to the pigs or chickens, it can be converted into concentrated foods which can be stored.

But dairy work is time consuming. It has to be done with reasonably fresh milk, of which the home dairyman will have limited quantities, and the trouble is that it takes almost as long to make butter and cheese from five gallons of milk as from fifty. So is it worth it? At today's subsidised prices, you get a low return on your labour time for making your own. But they will be superior, after practice, and greatly satisfying. Cheese

66

making is as absorbing a hobby as home wine making and, in many ways, very similar.

The main dairy products of interest to the backyarder are milk, cream, butter and cheese. Before we go on to consider production methods, it is as well we understand a little of what these products are.

What is milk?

Milk is an oil-water emulsion, the continuous phase being aqueous. (Richmond's *Dairy Chemistry*). That means that it consists of water with certain substances dissolved in it, plus little fat globules floating around in it. The proportions of the various constituents of milk vary considerably, but an average make-up of scientifically analysable materials is:
 — 87% water
 — 4% fat
 — 5% milk-sugar or lactose
 — 3% caesein
 — ½% albumen
(These last two are the proteins in milk)
In addition, there are bacteria, many of which arrive in the milk during and after milking, some of which are useful in the

human stomach or in the process of cheese making, but others of which are harmful. There are probably also loads of trace elements in milk, necessary for good nutrition but not easily analysed out, Milk is also known to be rich in calcium and in vitamins A and B. It also contains phosphorus, potassium, iron and vitamins C and D.

What is cream?

The little fat globules rise to the surface of settled milk and form cream. Eventually, the fat globules will settle at a certain concentration at the top of the milk and the familiar cream line appears. This process can be speeded up in a centrifugal separator. The butterfat content of cream is very variable — single cream is 18%, double 48% and the new third variety, sold as 'whipping cream' is 36%.

INTERIOR OF A DUTCH DAIRY.

What is butter?

There are basically two traditional ways of solidifying some of
the goodness out of milk so that it is storable and economically
transportable. The first way is the mechanical separation of the
fat globules from the water and the dissolved substances.
   Butter is formed by persuading virtually all the fat globules
in the milk to join up with each other into a continuous mass,
separated from the water and its dissolved constituents in the
milk. The persuasion that seems to work best is severe agitation
of cream at controlled temperatures, but the exact mechanism
by which it occurs is still a matter of dispute. Butter is made
up, very roughly, of:
   — 84% butterfat
   — 7% water
   — 7% unremoved buttermilk
   — 2% salt, added for flavour
The buttermilk, left behind after the butter separates out, is
made up of:
   — 90% water
   — 5% milk-sugar
   — 4% milk-proteins
   — 1% butterfat
So, butter is a good way of solidifying out the fat, but it leaves
behind the sugar and the proteins, not to mention most of the
minerals and the vitamins.

BUTTER-WORKER.

What is cheese?

The second method of solidifying is chemical rather than mechanical. If the acidity of milk is increased, the separate particles, particularly of milk protein, coagulate into a jellylike or custardlike substance called curd. A great deal of the water and a little of the dissolved substances separate out into a greenish watery liquid called whey. This separation takes place more easily at above sixty degrees F. The increase in acidity can be achieved by one of three methods.

1. Simply add an edible acid, such as lemon juice or vinegar, to the warm milk.

2. Let warm milk stand where it stays warm. Little organisms in the milk (*Streptococcus lactis*) convert the milk sugar, lactose into lactic acid and, hey presto, the milk curdles by itself. But if you leave it too long, another load of organisms take over and turn it into 'bad milk', so take care.

3. The conversion of lactose to lactic acid can also be achieved by using enzymes. Rennet is the usual one employed. It is taken from the stomachs of calves and processed commercially. Very small amounts of commercial rennet added to warm milk will produce curd.

To make cheese, the curd is separated from the whey. It is then eaten fresh as soft cheese, or pressed into moulds to dry out the whey and water completely thereby leaving it in a form which will keep for longer.

70

Cheeses vary enormously in composition but here is a rough guide:

%	Camembert	Cheddar
Water	45—51	27—34
Fat	21—30	26—30
Proteins	18—23	27—40
Lactic Acid	0.5	1.5

(from Richmond's *Dairy Chemistry*)

Cheese is valuable, not merely as a way of storing milk for winter, but in its own right as a concentrated protein food. The other great thing about cheese is that you can make so many delicious variations.

There are plenty of variables in cheese making:
— the method of increasing the acidity
— the temperature at which the curd is formed
— the acidity at which it is formed
— whether heat is used to produce further separation
— how much heat and for how long
— whether the cheese is pressed
— under what pressure and for how long
— whether the cheese is affected by other bacteria
We will have more to say about the use of these variables in re-creating famous traditional cheese later in the book.

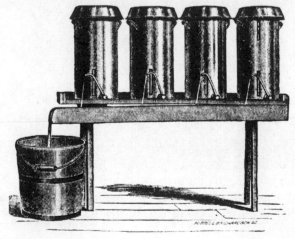

These, then, are the traditional forms of separation and they are the ones of most relevance to home dairying. But modern technology has discovered other interesting things to do with the milk.

Homogenised milk is milk, heated to reduce the surface tension of the fat globules, then forced through very small holes at high pressure to reduce the globules to a size at which they no longer rise.

Condensed milk is milk, mixed with sugar, then heated to around 235 degrees F and condensed in a vacuum.

Dried milk is concentrated milk. This is achieved by blowing hot air through the milk, thereby concentrating it. This concentrate is then placed in a rotating drum which pushes out the rest of the water. It becomes a brittle solid and is then made into powder. It is now also produced by a freeze drying process. Some dried milk is dried whole milk, some is dried skim milk.

Tinned cream contains a fairly low butterfat content. It is sterilised at 250 degrees F and should keep for a year.

Casein is sometimes separated out in large dairies. It is used to make nice white buttons and nowadays for that revolting artificial milk powder offered in sachets by mass-production caterers as a substitute for milk in coffee.

CURING-HOUSE, WHITESBORO' CHEESE FACTORY.

13 milk and cream

MILKING-PAIL. MILK-SIEVE.

After milking, the fresh milk should immediately be poured through a muslin cloth to strain out the flies and bits of dirt that inevitably collect in the milking bucket. It should then be refrigerated. The faster you bring down the temperature, the less danger there is from bacterial contamination. At that stage the milk is ready to drink.

Backyarders with Jerseys tend to refrigerate their milk in wide-topped containers. In twelve hours or so, the cream will have risen and can be scooped off. The milk left behind is still as good as silver-top and you have extra cream for butter making or just for spoiling yourselves.

The regular routine you must never forget is to wash out the utensils after each use: the milking bucket, the muslin cloth and the udder cloth. The simplest really effective routine is:

1. cold wash without soap (to remove milk-stone deposits)
2. warm wash in detergent, soap or detergent/steriliser
3. warm rinse, ideally with boiling water to sterilise
4. leave to dry without wiping

For safety, you can buy special dairy detergent/steriliser mixtures for the warm water, which cut out the need for a boiling rinse. Now this all sounds a terrible performance, and it is, but milk and milk utensils are, unfortunately, the ideal breeding ground for all sorts of very nasty micro-organisms. Where milk is concerned, this kind of extra care is essential.

Even with this routine, milk-stone deposits sometimes build

up on utensils as a sort of scale. You can buy special milk-stone removers from farm suppliers which will do the task.

Keeping milk

Goat's milk may develop off-flavours on keeping for over 2 to 3 days, even in the fridge, so we suggest either converting the surplus to soft cheese or freezing it.

A lot of books will tell you that goat's milk can be frozen, but cow's milk cannot. This is not so. Cow's milk can be frozen, but unlike goat's milk, it separates out into buttery fat globules and a thin liquid. This is no problem. When you thaw it out for using, whisk it up quickly and the milk recombines into its original consistency. So, whatever animal you keep, the extra milk of spring and early summer need not be converted to cheese or wasted; you can freeze it down for later.

Producing cream at home

Cow's milk separates into cream and skimmed milk fairly efficiently under the action of gravity alone. So does the milk of the Anglo-Nubian goat. But other goat's milk does not and the only effective way of producing fresh cream from goat's milk is by means of a separator. Frankly, this is a bore. But if you are a goatkeeper and sufficiently interested in cream, then try to pick up an old farm hand-separator by asking around at local farms or by advertising. The cheapest new separator still made is over £200 — out of the question.

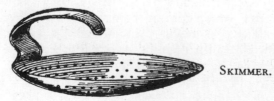

Skimmer.

74

N.Y. *Public Library Picture Collection*
SEVENTEENTH CENTURY GERMAN DAIRY.

Before you buy an old one, make sure that the tinning is un-broken and that all the cones are still there. The separator gives 99% of the available cream and you can adjust it for the thickness of cream you want. The trouble is that even the simplest machine has about 23 different parts, all metal, which need to be washed in soda and sterilised after each use. Even if you use a dairy disinfectant rather than boil the components, it is quite a chore when the amount of cream involved is small.

If you are the happy owner of a cow, then cream making is simple. There are two methods for home production.

The settling pan method. You need a big wide shallow pan (not plastic), and a cream scoop (a saucer will do), to draw off the cream once it has settled. Leave it for twelve to twenty four hours in a cool place and then skim off the cream. Start by breaking it off at the edges of the pan first. It comes off like a thick skin. This method gives you around 90% of the potential cream available. It needs very little equipment, but quite a bit of cool shelf or fridge space.

75

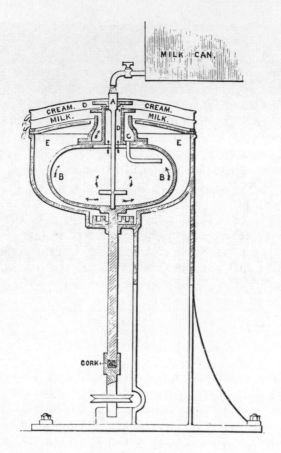

-Section of Separator.

The Syphon method. For this you need gallon jars, preferably caterers' salad cream jars because they have wide necks, and a wine syphon from a wine making kit or store. Slice an oblique cut at the end of the plastic pipe on the syphon. Keep the milk in your gallon jars for twenty four hours in a cool place. Then slide the syphon pipe down inside the edge of the jar past the cream layer to the bottom of the skim milk below. Syphon off the skim milk, finishing with the jar tilted over until you reach the visible cream line. You are left with cream in your jar. This method involves less storage space and the equipment is easier to get than a settling pan, but it is a bit more fiddly.

76

Of these methods we favour the syphon. The jars can be placed straight in the fridge. This speeds settling and gives a better-tasting cream than the settling pan method, because the large surface area of the milk in a pan tends to pick up taints. Also, it yields a larger percentage of the available cream.

The skim milk can be drunk fresh, but if you have enough whole milk left over for drinking, you won't want to bother with the thinner stuff. It is also an excellent food for feeding to calves, kids, chickens, pigs and other animals. Other than that, it is fine for turning into yogurt or skim milk cheese.

Clotted Cream

Clotted cream makes a pleasant change for 'cream teas' and the like. It keeps longer than fresh cream and so is better for butter making. It means that you can spread your butter making sessions out to once a week or more. You may find that clotted cream is the only foundation that works for butter making when the milk starts to get thin in the autumn. If you keep a goat, clotted cream is the only type you can make without a separator (and the yield is very low).

Stand the milk in a heatproof pan to a depth of 6 to 8 inches in a cool place for 24 hours. Transfer the pan gingerly to your stove and heat up gently to 150-170 F for use as clotted cream. It will form a ring at the edge of the pan, and the cream will split away. Let the pan stand to cool for 12 hours, then skim off the cream. It is not a lot of work, but it does take a long time.

14 butter making at home

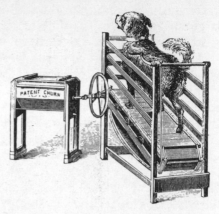

About one third of the World's milk production is reckoned to be turned into butter. The dairy industry of New Zealand, for instance, exists primarily to produce and export butter. In spite of the cheap subsidised price at which it can be bought, most backyarders will want to have a go at making their own. It will certainly be fresher and probably much tastier than the bought variety.

Here is a very simple butter making method from Jim Platts of Willingham, Cambridgeshire:

"We don't own a milking animal. We buy ordinary milk (not Gold Top) and take the cream off each bottle using a 5ml. plastic syringe. (Your doctor will throw away several of these each day). We put the cream in a two pound Kilner jar (the wide top makes it easy to get the butter out), and when it is half full, we stand it near the fire until it is just warm to the touch. Then we shake it by hand until pellets of butter form. We reckon usually on less than five minutes shaking to produce a yield of ½oz. of butter per pint of milk. We produce as much butter as we want from our weekly milk supply."

Now, for those who want it, here is a little more detail, and some ideas on scaling up the process to use kitchen machinery. There are three types of cream that can be used to make butter.

Fresh cream is fine if kept refrigerated prior to butter making. Alternatively, it can be kept by adding at least 8% by weight of

WORKING THE BUTTER.

salt — but this produces a very salty butter not to everyone's taste.

Clotted cream keeps well before the next butter making session and the taste of the butter obtained is particularly liked by some people.

Soured cream (or Ripened cream) makes a butter with a distinctive soured flavour. Making butter with soured cream is easier — it turns granular more quickly.

For a start, we suggest you have a go with fresh cream. If you experience difficulties, or feel like a bit of variety, try one of the other two.

Now the approach to butter making given in this book is rough and ready. It will produce butter quickly and simply on a regular basis, but it will not produce the ideal butter for keeping. This takes a lot more time and care and is, frankly, pretty unnecessary if you are making butter regularly for quick consumption. Besides which, this quick-method butter will freeze well, so doing it the old-fashioned way is no longer necessary anyway.

Equipment

You do not need to use a special butter churn for butter making but they are good for larger quantities. If you do not want to track down a churn, and if you are lucky enough to have an electric food mixer, then you have got nothing to worry about.

Butter can be made very quickly in the liquidiser or more sedately in the mixer bowl. Making it in the liquidiser, you tend to get a frothy butter granule pattern which congeals and from which it is very difficult to remove the buttermilk. So we suggest the mixer bowl fitted with the whisk attachment as the best equipment for butter making.

Getting the cream right for churning

Half the battle in butter making is starting off with cream at the right consistency and temperature. I may say that if your raw material is Jersey milk, you have got less to worry about anyway, because the large fat globules of this milk makes butter making generally pretty straightforward.

But even Jersey milk can get difficult as the lactation progresses. The earlier in the lactation, the easier and richer the butter. So, if you have a freezer, plan to concentrate your butter making accordingly.

Every book seems to say something different regarding temperature. The real point is that the higher the temperature (within reason), the faster the cream turns to butter. But also the less efficient the butter making in terms of loss of butterfat and the less firm-textured the butter and the more difficult it is to squeeze the buttermilk out of it. So the right temperature is a balance of opposing considerations. Since we are going for producing butter within an acceptable time (over 20 minutes, it gets tedious) and are not too concerned about efficiency or keeping properties, we suggest a cream temperature just below the temperature of your kitchen, but preferably within the range 57-65 degrees F. Within that, you should experiment. If you find difficulties, invest in a dairy thermometer and try to tighten up on temperature control.

The consistency is also important. The ideal is said to be 30% butterfat content. This is just under the content of commercial 'whipping cream', so buy some when you start off to use as a consistency guide.

The ideal acidity for butter making is apparently 0.5-0.6% but don't worry too much about that. Certainly it's true, however, that butter does come more easily if the cream smells slightly 'buttery'. Not 'cheesy', you understand — just slightly buttery.

Churning

Take a cupful of the thickest cream first and spoon it into the mixer, working at a slow speed (around 90 rpm). The bowl and head should be cold. Watch until patterns are made, rather like those that appear with whipped cream, then add another cupful of the cream and repeat until all the cream is in. Carry on mixing until the colour changes to golden yellow, and then until the wheat-sized granules of butter are formed.

The cream passes through three stages: soft whipped cream, stiff whipped cream and then a pretty sudden separation into yellow granules of butter slurping around in white buttermilk. This last change is accompanied usually by a change in the sound coming from the mixer-bowl.

If you wish, you can stop at this stage. You have made butter.

ECCENTRIC CHURN.

81

Just squeeze the buttermilk out in a muslin bag. This will remove most of it and give you butter which is fine for the next week's use. If you want something you can store, refrigerated, for months, you had better wash it and work it well.

Purifying the butter

The traditional next stage is to add 'break water', which is cool water that serves to keep the grains separate and round. Keep the mixer on slow as you do this until you get as near as possible to the ideal of perfectly round 1/8inch diameter granules, all quite separate from one another. But don't worry if you can't achieve this, it is only a way of expelling more buttermilk and so making the butter last longer.

Alternatively, now is the stage to pour off the pure undiluted buttermilk. If cholesterol consciousness goes much further, it would not surprise me to see butter considered the by-product of the making of buttermilk, rather than the other way round. After all, the proteins, milk sugars and other dissolved nutrients of milk nearly all go into the buttermilk during the separation process — surely much to be preferred to a slimy yellow mass of almost pure animal fat!

Seriously, the buttermilk is worth saving. Drink it fresh, use it for bread-making instead of water, or make cultured buttermilk, a fine-tasting product which you pay heavily for in the shops. Add one part of ready-cultured buttermilk, either from a previous load or shop-bought, to eight parts of new buttermilk. Let it thicken in a warm place and then store it in a fridge. Alternatively, you can use a commercial starter or lactic acid starter, which has the advantage of keeping longer.

Back to the butter now. It can at this stage be washed to get rid of buttermilk. Keep rinsing until the washing water stays clear (5-10 washes should be plenty).

After the final straining out, sprinkle on some salt if you like. We suggest one teaspoonful per pound of butter. For the really professional touch, add some annatto butter colour — particularly if the butter looks less golden-yellow than the stuff which you buy in the shop and are used to. It probably will do, except in the very best grazing periods of early summer. See the cheese section for how to make your own colouring, and for addresses of annatto suppliers.

Working the butter

If you want the butter to keep, its water content should be less than 16%. Commercial butter is more likely to be around 12%. To remove the water, work the butter. The simplest method is to start with is to use a cutting-board and a strong wide-edged knife or spatula. Squeeze down the butter onto the board, forming a thin layer. If you slope the board (say, on the edge of the draining board), the water you squeeze out will run off immediately. Fold it over and repeat the squeezing down, without rubbing or smearing the butter. Keep it fairly cool during working and keep going until you really are not getting much more out.

The traditional tool for working butter is a pair of 'Scotch hands' — flat wooden paddles. These are available from Smallholding and Self-Sufficiency Supplies. Remember to press, not rub, the butter, otherwise you'll get in a terrible mess and do the butter no good at all. Remember too that if the butter is too soft and flabby to start with, either because you didn't wash it or because it is too warm, don't work it in this state — it's pointless. Working will only expel water if the butter is firm enough to squeeze it out.

Care of butter making equipment

Right through this dairy products section of the book, we keep going back to the importance of cleaning utensils properly. It is very easy just to treat butter making utensils along with the rest of the kitchen washing up. That just won't do. Unless you take extra care, sooner or later you will get 'off flavours' in the butter.

Either do the dairy disinfectant routine described in the last chapter, or scald the equipment each time after use with boiling water. Some people advise rubbing Scotch hands with salt to stop butter sticking, but the snag is that if you over-do it, the wood gets very rough. If butter sticks, boil the Scotch hands in soda for a while to release the butter trapped in the wood.

NOVEL METHOD OF CHURNING.

Storing butter in brine

If you do not possess a freezer and do not mind fairly salty butter, here is an old idea for preserving butter. Make the butter in the normal way except for the addition of 1 oz of sugar for every 3 lb of butter. Butter for storing should be made the long way with as little buttermilk or water left in it as possible.

The brine consists of 4 lb of salt, ½ lb demerara (or soft brown) sugar and 1 oz saltpetre boiled in a gallon of water. When cool, the brine is strained through muslin into a clean crock or enamelled bucket. The butter is wrapped in grease-proof paper when made and then submerged. To keep it from floating, a clean slate or tile should be placed above the butter. It is ready for use straight from the brine.

Butter making problems

There is really only one big problem that occurs in butter making and that is when for one of many reasons, the butter

just does not 'come' in 20-30 minutes. If 40 minutes have passed and you have still got white cream, it could be for one of the following reasons:
— Your cream is too warm or too cold. Sometimes in winter it really needs to go up to 65-70 degrees F.
— Your equipment may not be really clean and sterile. This may cause what is known as 'ropy fermentation'.
— The animal may be almost through her lactation, in which case, butter is harder to make.
Here are a couple of other tips we have heard but not fully tested out. If goat's cream gives you trouble getting too big a granule size, add a cupful of water per one and a quarter pints of cream as soon as the granules appear. Also, it is said that thinner cream is better in summer and thicker cream is better in winter.

N.Y. Public Library Picture Collection

DAIRY TOOLS.

15 cheese production at home

<small>HOOP FOR FLAT CHEESE.</small>

As children of the super-industrial state we have been condit-
ioned to the standardised artificially cultured cheese of uniform
flavour and consistency. But just consider, before you start
home production, how things were before the Industrial Revol-
ution. Each cheese depended upon the quality of pasture, both
in terms of geography and season; it depended upon the size of
the farm and whether or not the cheese was made for sale (in
which case, even then, it had to be more uniform). This infinite
variety was at the basis of the many 'standard' varieties of cheese
produced now in the rigidly controlled conditions to ape the
way the cheese came out naturally from certain farms.

For instance, 'moorland' cheese was only made in Spring,
when the cows emerged from winter quarters and grazed on the
young scented herbs and grasses of the moors. This diet cleared
their blood and they gave a beautiful casein-rich milk which in
turn produced a beautiful, light-textured, blue cheese. The
moorland people would never have dreamed of making this
cheese at any other time of the year — but a cheese factory
has to keep going throughout the year.

Take another case. The blue-veined Stilton depends for this
effect on the action of the mould *Penicillium roqueforti*. The
cheese rooms in the Stilton district of Leicestershire were well
infected with this mould and the process carried on naturally.
Nowadays, the milk for Stilton making is heated to destroy the
existing organisms in the milk. *Penecillium roqueforti* is then

injected in measured quantities to ape the conditions of these old Leicestershire cheese rooms.

The point is that the characteristics of each local type of cheese depended originally upon the peculiarities of the local conditions to a very large degree. We could, in this book, give you a great stream of recipes telling you how to mimic the local cheeses of every district of England, France and the rest of the World. Instead we have chosen to introduce the basic processes common to most cheese making, plus some variations developed in certain areas and worth a try in your dairy. We hope this will spur you on to produce unique individual cheese which varies over the year and with your experiments. Maybe, after a while, you will decide to stick to a particular variant which pleases you best.

But before we go into the processes of cheese making, we must give you a very definite warning. Making cream, butter or yogurt is a pretty simple affair; making soft cheeses is also not too difficult, but making hard cheeses is a different ball game. Most backyarders we know admit to failure rates of between 50% and 90%, even after a year or so of trying. Some have given up altogether.

If your rationale for backyard dairying is primarily the pleasure of producing superior food from your own resources then you may do well to steer clear of hard cheese making. In all honesty, we have yet to taste a backyarder's cheese that tasted better than what is normally available in the shops. That is rather a sad admission, but there it is.

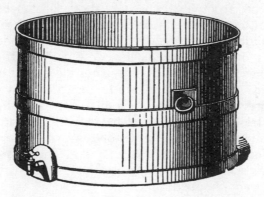

CHEESE-TUB.

87

Nevertheless most of you, we are sure, will want to have a try at it, or at least to understand its complications and skills enough to decide to steer clear of it, so here we go.

In chapter 12, we explained the basic chemistry of turning milk into cheese. Now we will proceed to some practical recipes for making cheeses. You will remember that there were three ways of creating the necessary acidity for curd formation. Here are basic recipes for each way:

Adding acid. This is the method used for cheese making in India. It produces a soft, rubbery cheese called 'Panir' which is chopped into 2" x 2" x ¼" blocks, fried or added to curries. Panir is also used to make certain Indian sweets.

Heat one and a half pints of milk, stirring, to boiling point. Remove it from the heat and add quarter of a teaspoonful of tartaric acid (Cream of Tartar) dissolved in a cup of hot water. Stir until the whole of the milk curdles. Leave it for fifteen minutes covered and then strain it through muslin and squeeze out all the whey. At this stage it is called 'Chenna'. To make Panir, keep it in muslin, tied tight and place it under a 5 lb weight for two hours. For more details on how to use this in Indian cookery, get hold of Mrs Balbir Singh's book *Indian Cookery*.

Letting the milk sour. This is the method behind all the various so-called 'lactic' cheeses (meaning cheeses soured by lactic acid rather than rennet). It is a simple method for the home production of very pleasant soft cream cheese. This can be made from ordinary soured whole milk or from soured buttermilk, or from yogurt. Any of these substances, as long as they are thick, can be drained in a muslin or cheesecloth bag to produce a form of soft cheese. It helps to open the bag after half and hour and remix the contents. Repeat this again after another hour. By doing this, a more even, creamy cheese is produced. After that, there are further possible variations. You can press the cheese between plates for an hour or so, or add some fresh cream until the cheese works up into a fine pat of cream cheese. Salt the cheese to your taste and maybe end up by adding some form of extra flavour. The possibilities here are endless. Try starting with chives, wine (stirred in), orange slices or chopped nuts.

The cheese made by draining yogurt overnight is very pop-

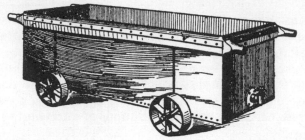

RECTANGULAR CHEESE-VAT.

ular in Germany under the name 'Quark', and is quite delicious used as we would use whipped cream on fruit. After draining, the quark is whipped up in a bowl until really smooth. The addition of fresh milk or cream can help the texture and flavour but 'low-fat' quark is generally preferred these days. Finish off with a little salt to taste.

Adding rennet. This is the method for all hard and semi-hard and for a lot of the better soft cheeses. This is where cheese making gets interesting but rather complex. Let's start with a simple version called 'Neufchatel'.

Use 8 pints of any milk (as long as it is not 'off'). Heat to 86 degrees F and add 2 drops of rennet dissolved in a tablespoonful of water. Leave it for 24 hours in a airing cupboard, then ladle the set curd into a muslin cloth and hang for 8 hours or so until it stops dripping. Salt it and it is ready.

All these soft cheeses are pretty bland, so it is an idea to add highly-flavoured things to them. Anything will work except unpeeled or soft fresh fruit which can contain organisms that make the cheese go 'off' quickly. Our favourites are:
- honey and chopped nuts
- red peppers
- mixed fresh herbs, particularly dill and fennel
- tinned pineapple chunks
- crispy bacon bits

One rather special and very popular soft cheese is the American-style 'cottage' cheese with its characteristic lumpy consistency. This is somewhat trickier to make.

89

Cottage Cheese

Make curd as above and when jelly-like, showing clear whey as
the surface is broken by your finger, cut the curd gently into
½" cubes. Raise the temperature very slowly over an hour to
95-100 degrees F, then leave the curd in the whey for another
20 minutes. Drain the curds through muslin, then wash them
in cold water and salt to taste. The idea is to shrink and set the
cubes of fresh curd into the familiar lumps of cottage cheese.

Hard Cheeses

They are made by starting off as for soft cheeses — separating
curds from whey. Usually rennet is the coagulating agent used.
But certain treatments follow in order to improve separation
from the whey and firming of the curd. The firm curds are then
packed into moulds and pressed under weights to further com-
press and firm the curd. Finally the curds are allowed to stand
and ferment over a period of time so that they take on the
characteristic flavour of hard cheese. Although there are a great
many variations, the making of hard cheeses generally follow
this basic ten step pattern.

1. Ripening the milk. This is a delicate process and one where
you will need to experiment according to your own conditions
and tastes. If you let the milk sour too much, you may end up
with a curd which is an acid curd/rennet curd mixture, difficult
to ripen. If you use milk too fresh, the acidity is inadequate for
the rennet to have the desired effect.

The traditional farmhouse solution is to let the evening milk
ripen overnight at a temperature of 50-60 degrees F. In the
morning, warm it and stir the cream back into the milk. Add
the fresh morning milk and your cheese-making milk should be
about right. You can, alternatively, add commercial starter
(lactic acid culture) to fresh milk and let it stand in the warm
to ripen it. Whatever you do, the milk should still taste sweet.

2. Colouring the milk. This is optional but red annatto is
added to certain cheeses. This is up to you. You could experiment with using saffron or marigold petals or the juice from
pressed grated carrots. This is what used to happen before
annatto was imported.

3. Coagulating the milk with rennet. Extract of rennet is stirred in at around 85-90 degrees F for most cheeses. Some cheeses are rennetted as high as 105 degrees F. The usual practice is to add the required amount of rennet, diluted 1:6 with warm water. So long as the acidity is right, this will produce a curd hard enough to hold the fat globules together without being tough. Rennet curd does shrink naturally and this helps to expel the whey. The higher the temperature and the acidity, the faster this shrinking takes place. If there is too much acid in the milk, it will eventually interfere with the shrinkage process and you end up with a nasty wet acid mess. Make sure that you stir the rennet well in and then leave the milk undisturbed at 85-90 degrees F until it has set to the consistency of junket. This may take anything from 15 minutes to an hour. The usual way to test curd is to poke your index finger under the surface, then lift. If the surface breaks cleanly, then it is ready.

4. Breaking the curd. You can leave the curd unbroken to drain, but the process is speeded up if you break the curd into pieces. You get better results if you keep the pieces of curd even in size. Most English varieties of cheese are made with ¼ to ½ inch cubes. The illustration shows one way of getting fairly even cutting with an ordinary kitchen knife.

The difficulty of producing cubes of even size right the way down can otherwise be achieved using a curd knife which is obtainable from either Self-Sufficiency and Smallholding Supplies or Small-Scale Supplies.

5. Treating the curd. The purpose of this stage is to firm the curd more effectively by heating it in the whey. It is also to improve the drainage of the whey from the curd by keeping the curd pieces gently moving. You can carry out both treatments at the same time, or separately. If you break the curd with a knife, it is perhaps better to start by stirring the curd gently with your hand for about 15 minutes. This way, you can check that the pieces are all fairly even and break up those that need it. Try to stir as gently as will keep the curd pieces from sticking together. After this, you can heat the pan very gently to 100 degrees F over about an hour, stirring from time to time to keep the pieces from sticking. The curd pieces should then have firmed up sufficiently to fall apart in your hand without any squeezing. If you keep them at 102 degrees F for another hour,

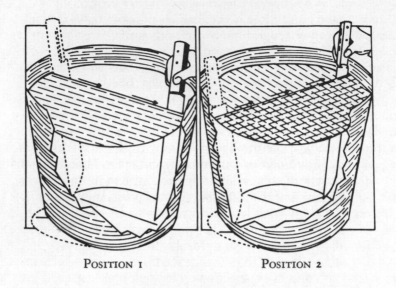

POSITION 1   POSITION 2

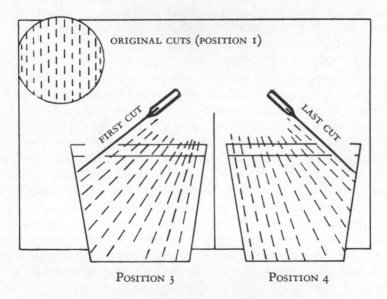

ORIGINAL CUTS (POSITION 1)

FIRST CUT   LAST CUT

POSITION 3   POSITION 4

FOLLOW YOUR ORIGINAL CUTS AS NEARLY AS POSSIBLE, HOLDING KNIFE
AT ANGLE AS IN POSITION NO. 3—THEN AS IN POSITION NO. 4.

this will firm up the pieces even more so that a handful of pieces will shake apart after being pressed together in your hand. This is the degree of separation you should aim at as a beginner before you go on to drain the curds. Later you may wish to shortcut this rather long process. Do not leave the curds and whey together too long because the acidity starts to rise and the curds go wet and unmanageable.

6. *Draining off the whey.* This is most simply done by pouring the curds through muslin. The only skill is in deciding when the curds are ready to be drained.

7. *Treating the drained curd.* This very important stage determines to a large extent the final texture and ripeness of your cheese. The degree to which you further break up the curd at this stage, by cutting or squeezing, determines its texture. If you like crumbly cheese, keep the pieces loose. If you like it more resilient, squeeze the curds together into more solid blocks. Before you do any of this, however, you must salt the cheese. One tablespoon per sixteen pint load is a good starting guide. The main point of adding salt at this stage is to stop the action of the acid-bearing baccillae. For certain cheeses, you chop or mill the curd at this stage as well.

8. *Forming the pieces into cheese.* At home, this is most easily done by forming a 'bandage' of cheesecloth around a cake-shaped mass of curds about six inches across, and pinning the bandage in place. With a couple of thicknesses of cheesecloth above and below the 'cake', it is ready for pressing. Alternatively, you can make your own wooden cheese mould out of wooden storage jars, available from kitchen supply shops, with a few holes drilled in the sides and bottom to drain out the whey. This kind of mould, however, will not stand much pressure. You could also try making cheese moulds from old stainless steel saucepans or large food cans with holes drilled in them. Whatever you use to contain the curds, make sure that they are packed in well, with no cracks extending into the centre of the cheese. Again, the shape and size of the mould also depends upon which regional cheese you are trying to copy.

9. *Pressing the cheese.* If you can get hold of an old cheese press, so much the better. If not, any simple device will do, providing it gives an even downward pressure onto the cheese,

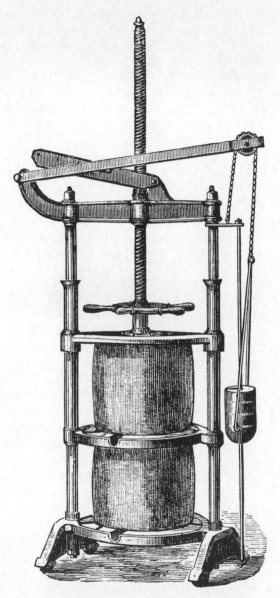

CHESHIRE CHEESE-PRESS.

which continues to be even when the cheese shrinks down. At its simplest, this can be an empty can which slides down into a can mould, with two bricks placed on the inner can. The usual methods of pressing involve an initial pressing (you can start with two bricks), then an upending of the cheese and a second pressing at a higher pressure (try four bricks). The period of pressing can also vary, but we suggest you start with 12 hours at each pressure. This should be fine for small 6" cheeses but naturally the cheese factories these days press their products under pressures of several tons to speed up the process.

10. Ripening the cheese. The fermentation process that takes place at this stage can give the cheese a fine mellow flavour or can ruin it completely. Be warned that, for no apparent reason, your cheese may dry into a pile of acid crumbs or it may stick to the shelf in a slimy mess, or it may even roll off the shelf completely — propelled by the gas of its own fermentation.

The cheese should be transferred to a cool but frost-free cellar or pantry. Beforehand, you may choose to protect the cheese once the rind is dry by painting on a layer of paraffin wax, heated first in a double saucepan. You can also use lard or cornflour. Ripening cheese should be turned regularly every day for the first few days, then every other day until the cheese is hard and ready — usually in three to four weeks. You can

taste it at intervals, if you wish, by taking out small pieces with an apple corer. Make sure you put back the outer plug, and seal the wound again with wax, lard or cornflour. Most cheeses are best left until at least two months old.

Ringing some changes

    — add butter to the drained curds
    — try 'green cheese', that is, cheese freshly pressed but not ripened
    — put leaves inside the press before you fill it with curds: dock leaves, nettles, vine leaves are some which are used to wrap cheese
    — try letting some of your cheeses go 'blue' by leaving them up to six months. Depending upon the organisms present in your pantry, interesting things may start to happen.
    — when salting, grate some cheese of the desired type into the curds. This sets up a good bacterial growth for the ripening stage.

Copying well-known cheeses

The one that you will perhaps most want to copy, Cheddar, is one of the least suited to small-scale production. But it is possible and Self-Sufficiency and Smallholding Supplies can sell you the proper moulds, though even the recipe that they supply starts off with "Take six gallons of whole milk..."!

Still the best source for recipes of virtually all the well-known cheeses is John Ehle's *Cheeses and Wines of England and France* published by Harper and Row.

If you are using Jersey milk, we suggest you start with the softer 'hard' cheeses first and ones that require little handling

GERVAIS CHEESE MOLD.

N.Y. Public Library
Picture Collection

96

of the curd. Colwick, Cambridge, Coulommier and Pont l'Eveque are all suitable. Work up from there.

If you are a vegetarian, you may want to try using casein rennet rather than the usual stuff. We found this recipe for making your own:

Set six pints of milk as for clotted cream. Skim the cream off and add two tablespoons of vinegar to the skimmed milk; then heat it until it curds thoroughly. Wash the curd in water three or four times, kneading thoroughly. Dry the curd and powder it. This powder is almost pure casein.

Alternatively, if you are very adventurous, you could try using plant juices to replace rennet. It is said that the juices of the nettle, sorrel and ladies' bedstraw will work, but we have not yet tried this one out.

CURD-MILL.

16 yogurt production at home

Fifteen years ago, if you had done a survey, I doubt whether 10% of the population would have known what yogurt was. All that has changed now. Big food manufacturers recognised its potential as a bland, creamy-textured food, eminently suitable for flavouring with lots of different sweetened fruit flavours and also having a 'health' image. Yogurt must have been one of the biggest food success stories since baked beans.

Needless to say, the traditional product is rather different from what we get from our supermarkets — as you will know if you have ever tried it in Greece or the Middle East. Traditional unboosted, unsweetened yogurt has been popular throughout Northern and Central Europe and Western Asia, but is supposed to have originated in Bulgaria. There are many local variants, from *Skyr* in Iceland, through *Dahi* in India, *Taette* in Scandinavia, *Leben* in Egypt, to *Mazun* in Armenia.

Traditional yogurt is milk coagulated by the use of added cultures into a soft curd. In hot countries, culturing is a way of preserving milk for slightly longer in a palatable form. In addition, it has had ascribed to it various therapeutic properties, mainly, it appears, because of the unusual health and longevity records of people in certain countries where it is eaten regularly. Recent research suggests that yogurt might be beneficial in the treatment of migranes. Who knows — it's certainly more pleasant than pills.

The cultures normally used are *Lactobacillus bulgaricus* and

Streptococcus thermophilus, and the traditional idea is that regular consumption of yogurt establishes a colony of these micro-organisms in the intestines and that they assist digestion. However, some recent research suggests that these particular organisms are not, in fact, able to establish permanent colonies but that they just help things along on their way through. This means that you have to keep eating yogurt regularly for it to be beneficial.

There is more to come. Milk-curdling cultures have recently been discovered which occur both in mother's milk and in the stools of infants. One is *Lactobacillus acidophilus.* This also occurs naturally in cow's milk. This culture is known to be particularly beneficial in that when established in the lower intestine, it produces conditions unfavourable to the growth of harmful putrefactive bacteria. It also aids the breakdown of food particles and the synthesis of vitamins B and K.

WATER-SEALED CAN (c. 1877).

The trouble is that these friendly intestine dwellers are killed off when you take penecillin and most other antibiotic or sulphur drugs. So, if you do have to take these kinds of drugs, it might be an idea to follow them up with a few doses of yogurt made with *acidophilus.* Certain strains of *acidophilus* have now been developed which are resistant to a range of antibiotics and which can be taken in powder form or made into yogurt. The stuff to buy is Enpac, made by Aplin and Barrett Limited, and available from chemists, though they will probably have to order it specially for you. Be warned, it's not cheap. The last batch cost me £1.30. Also note that it works best on skimmed milk. You may, alternatively, be able to find *acidophilus* yogurt in a good wholefood shop to use as a starter.

Whether you use *acidophilus* or the more usual cultures, the principles are the same. You add a small quantity of culture, or milk from a previous batch containing culture, to milk, and then you incubate the cultured milk until the culture has thoroughly multiplied and coagulated the milk to the desired consistency.

Any plain bought yogurt will work as a starter. Do not be fooled by the labels in healthfood stores saying 'live yogurt'. So far as we know, commercially sold natural yogurt, at least in the United Kingdom, is all live.

There are many variations possible on this basic method. If you are using some shop-bought yogurt as a starter, add it in at about one tablespoonful to the pint, mix it in well, and hold it for 3 to 4 hours at 70-105 degrees F. If you buy culture, stick· to the supplier's directions on time and temperature.

Dahi is based on buffalo's milk, which is very rich. Much of Middle Eastern yogurt is based on goat's milk, which makes a

A NORMAN DAIRY.

100

very sharp, rather thin, yogurt and tastes beautiful. But my alltime favourite is the *Yaorti* that you get in Greece, made from ewe's milk. It has a beautiful, light, yellow, creamy crust, and is firm and jelly-like. Served with a little clear honey on top, it is out of this world. You can produce a copy based on cow's or goat's milk without too much difficulty. The principle is to incubate the milk in a fairly shallow, vertically-sided dish with heat fully surrounding it. We use either a pate dish or several

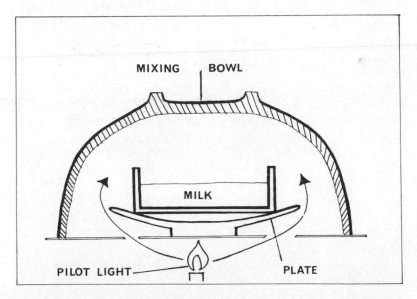

MIXING BOWL

MILK

PILOT LIGHT

PLATE

plastic commercial cottage cheese pots. Place them above the pilot light of a gas stove, covered by a very large mixing bowl and under this arrangement, the cream settles into the right consistency of crust and the plate stops the base of the pate dish getting too warm and spoiling that layer of yogurt. The milk is heated and cultured in the normal way and left overnight. This produces an unbeatable Greek-style yogurt.

If you or your children are hooked on supermarket yogurt, that's not difficult to copy either. Add dried skim milk or dried whole milk (whatever it says in the 'ingredients' printed on your favourite brand) to the milk you use. You will need to experiment to get the right final consistency. For flavouring, use a good full-fruit jam and mix it up well. This way you can make a 'modern' yogurt for under half the supermarket price. Nevertheless, supermarket yogurt is year by year getting harder to copy. The ingredients on the last pot I bought were: 'concentrated skimmed milk, sugar, fruit, cream, cornflour, gelatine, colour, flavouring'. No mention of fresh milk or culture any more, so perhaps commercial fruit yogurt is thickened with cornflour and gelatine nowadays and no culture is needed?

Keeping a batch of cultured milk at around 80-90 degrees F for a few hours needs a bit of care. Apart from the pilot light method described above, you can use a vacuum flask, the airing cupboard, thick insulation or buy an electrically heated 'yogurt maker'. The simplest method is to use a big vacuum flask. Warm it first. Heat the cultured milk in a saucepan to blood heat and put it in the flask. We leave ours overnight, but that is out of laziness. It is less acid if you cool it right down once the culture has formed a complete curd — usually in 3 or 4 hours.

If you like your yogurt jelly-like in individual pots, then embed the pots in a box of polystyrene beads (sold for making 'sack' chairs) or vermiculite roof insulation chips. Cover the filled box with something equally insulating like polystyrene sheeting, or a blanket. Do not let the polystyrene beads get in the yogurt — they taste most peculiar, and cannot be good for you.

We do not think that the expensive electrically heated 'automatic' yogurt makers are really worth the expense, especially when you can do the job so simply without them.

One more point on hygiene. In theory, you can keep going indefinitely with home-made yogurt, keeping a little back each time to start the next batch. The problem is that the incubation can encourage the growth of other, less desirable organisms. If you intend to keep your own culture going, be sure only to add the culture to milk that has just previously been boiled and then cooled to blood heat in a covered container.

Frankly, we don't like the taste of boiled milk, nor is it nutritionally as good for you — so we prefer to keep the culture going only two or three times and then replace it with a pot of shop-bought plain yogurt.

17 various other dairy products

There will be times when your cow, goat or ewe produces more milk than you can immediately think what to do with. Here are some ideas to help you out.

Sweet Puddings

Junket is just milk set with rennet (the basis of cheese) eaten fresh. The rennet used is in the form of 'junket tablets' and can be bought from good grocers. Follow the directions on the packet, but basically all you have to do is heat the milk to around 90 degrees F, add the tablet and leave it to set.

The traditional English flavouring for junket was rum, but powdered nutmeg or cinnamon on top also helps. Personally, I find junket a very dull dish compared to either yogurt or blancmange.

Blancmange. The addition of two and a half tablespoons of cornflour to a pint of milk, which is then boiled and cooled to set, turns the milk into a very acceptable pudding. It can be flavoured a hundred ways, but almond essence and chocolate are my favourites.

Ice Cream

The home production of ice cream is not as widely accepted in

Britain as it is in the USA, partly because our summers are not as hot and partly because so few of us have yet discovered the vast superiority of ice cream made in a special ice cream maker rather than in the ice tray of a fridge. Unless it is stirred vigorously *whilst* freezing, ice cream sets unevenly and forms nasty crystals. An ice cream maker is just a double saucepan with a built-in stirrer. In the outer pan goes freezing brine, which you make by dissolving as much salt as the water will take, then freezing it in the fridge ice-tray or your deep-freeze. The advantage of brine is that it freezes at a colder temperature than pure water. You then put your pre-cooled ice cream mixture in the inner pan and stir it until it forms a thick creamy genuine ice cream.

Cooking Fat

Ghee is clarified butter as prepared in India for use as cooking oil. It keeps for months, if not years, even in their climate. The purpose of clarification is to separate out the milk-proteins

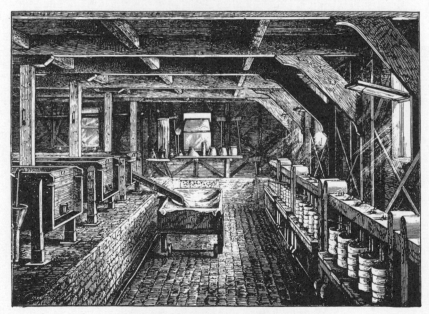

LONGFORD FACTORY (INTERIOR).

from the butterfat; it is these proteins that make butter go off. What you are left with is almost pure butterfat (plus salt if you added it in the first place).

The basic way to make *ghee* is to cook butter on a slow heat and allow all the water to boil out. At the point when the sediment begins to brown, take it off the heat. Pour off the clear liquid through muslin and store it in closed jars. The trouble is that if you leave it too long, the sediment burns and gives a nasty taste to the *ghee*.

Here is a safer method. Heat the butter on a medium heat until the froth has all been stirred in. Cool the butter and keep it in a fridge for three or four hours. You will see that the *ghee* sets in a thick layer on the top with a protein sediment underneath. Take off the *ghee* carefully with a knife and boil it up again for one or two minutes. This produces crystal-clear *ghee*.

Use for Whey

Whey cheese. Carrying on the principle of not wasting any of that good milk, here is a Norwegian recipe for using up the whey from cheese making. Boil it slowly until it has evaporated to a creamy consistency. Then stir it well and keep boiling until

it is really pastelike. Spoon this into a greased bowl to cool. Then tip it out onto a plate for serving. It is brown, highly nutritious and much sought after in Norway. You may find that that it takes a bit more getting used to.

Blaand. "Pour the whey into an oak cask and leave it till it reaches the fermenting, sparkling stage". (Maybe those of us without oak casks could get away with using a plastic wine making bucket?) "It is delicious and sparkles like champagne.

After a while it goes flat — but keep it at perfection by regular addition of fresh whey. *Blaand* used to be in common use in every Shetland cottage." (from *Recipes of Scotland* by F. Marian McNeill, published by Albyn.)

A couple of Eastern oddballs

Khoa. This dried whole fresh milk is the basic ingredient of Indian milk sweets. Boil a pint of good creamy milk in a thick-bottomed pan on a fast heat. Stir gently in the initial stages — just to prevent it from boiling over. Then, when it starts to thicken, stir more vigorously to keep it from burning. Eventually, in about 20 to 25 minutes, you end up with a single lump in the bottom of the pan. That is *khoa*. Remove the pan from the heat, but keep stirring till it stops sizzling.

Kefir. This is rather peculiar fermented milk, which is alcoholic and gaseous. You have to get hold of *Kefir* grains, also known as 'yogurt making plants' which go white and gelatinous in milk. Use skim milk at about 60 degrees F and shake the grains and milk from time to time. The grains will have done their work after about 12 hours. They can be drained off, dried and stored. If you cork the milk for another 12 hours, you get *Kefir*. Strange stuff, but popular amongst the wandering tribes of Central Asia.

106

references

Recommended works of reference on husbandry

D. Mackenzie, *Goat Husbandry*, Faber.
Dirk van Loon, *The Family Cow*, Garden Way Publishing, USA.
Ann Williams, *Backyard Sheep Farming*, Prism Press.
Ann Williams, *Backyard Pig Farming*, Prism Press.
— all are described in the text and are the best reference works in their fields.

Newman Turner, *Herdsmanship*, Faber.
— the older standard work for an organic approach to keeping cows. If you spot anything by the author in a secondhand bookshop, buy it.

K.Russell, *Principles of Dairy Farming*, Illiffe.
— the commercial approach.

The new crop of books on goatkeeping.

E.Downing, *Keeping Goats*, Pelham.
— amusingly written, but expensive for what is a pretty basic book.

H.E.Jeffrey, *Goats*, Cassell.
— humourless style with lots of pictures of show champions — seems geared to school goatkeeping.

J.Salmon, *The Goatkeeper's Guide*, David and Charles.
— the best of the new crop, but still no match as a reference
book for Mackenzie.

L.Hetherington, *Home Goatkeeping*, EP.
— sketchy.

L.Hetherington, *All About Goats*, Farming Press.
— a third of this is by the TV Vet and is useful stuff, but again
lots of pictures of champions.

J.Belanger, *Raising Goats the Modern Way*, Garden Way Publishing, USA.
— the only one with a homesteader approach, but geared, of
course, to the USA.

All these publishers (bar the last) have suddenly cottoned on
to the upsurge of interest in goatkeeping and have presumably
commissioned 'well-known' goatkeepers to write them a book.
The result, generally, is a bias towards herds and towards the
business of showing goats.

Recommended reference on cheesemaking

John Ehle, *The Cheeses and Wines of England and France*,
Harper and Row.

New books on dairy products

Susan Ogilvy, *Making Cheeses*, Batsford.
— very nicely produced practical book, but not that many
recipes.

M.Black, *Home-made Butter, Cheese and Yogurt*, EP.
— covers similar ground to the latter half of this book, but ex-
pands out into lists of flavouring for yogurt etc.

Technical works for the serious student

Kirschgesner, *Nutrition and the Composition of Milk*, Crosby
Lockwood.

Judkins, *Milk Production and Processing*, John Wiley.
— very informative US book about modern advanced farming.

Elspeth Huxley, *Brave New Victuals*, Chatto and Windus.
— the other side of the story.

Richmond's Dairy Chemistry, Griffen.

Advances in Cheese Technology, FAO.

The History of Dairying

Alan Jenkins, *Drinka Pinta*, Heinemann.
— the story of the Milk Marketing Board.

Val Cheke, *The Story of Cheese Making in Britain*, Routledge.

Dorothy Hartley, *Food in England*, Macdonald.

More general books on 'Back to the Land'

William Cobbett, *Cottage Economy*, (1850)
— the original back-to-land book.

John Seymour, *The Complete Book of Self-Sufficiency*, Faber.
— beautifully produced, but the claim to completeness is rash.

John and Sally Seymour, *The Fat of the Land*, Faber.
— in my view, the best of John Seymour's books — it tells how they got started.

Ken Kern, *The Owner-Built Homestead*, Scribners, USA and distributed in UK by Prism Press.
— terrific on the building side of self-sufficiency.

Useful Addresses

The Dexter Cattle Society,
'Lomond', Seckington Lane, Newton Regis, Tamworth, Staffs.

Countrywide Livestock Limited,
(for British Milksheep)
Eastrip House, Colerne, Chippenham, Wiltshire.

The British Goat Society,
Rougham, Bury St. Edmunds, Suffolk.
— this is the national umbrella organisation — they publish some useful booklets and also sell various treatments etc.

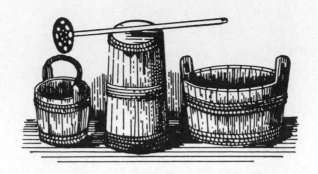

Suppliers

Self-Sufficiency and Smallholding Supplies,
The Old Palace, Priory Road, Wells, Somerset.
Tel: 0749 72127.

— Their catalogue is a cornucopia of amazing artefacts. The
section on dairy equipment includes electric butter churns,
three-legged stools, udder cream, rennet, various styles of
cheese press and much more. Then there are the livestock and
cow care sections which are equally comprehensive.

Small-Scale Supplies,
Widdington, Saffron Walden, Essex.
Tel: 0799 40922.

— This is the mail order offshoot of *Practical Self-Sufficiency*
magazine (at the same address). At present, they concentrate
mainly on relevant books, of which they have a fine selection.
They will soon, however, be expanding their coverage to include a full range of dairy equipment: churns, cheese moulds,
cheese presses, curd knives, rennet, starter cultures plus all-in
beginners' kits for buttermaking and hard cheesemaking.

The above two suppliers are trying to serve the special needs
of backyarders and as such are in the best position to provide
the full range of what's needed. There are, in addition, however,
specialist suppliers of particular ranges, some of whom we list

below. But don't be surprised if the quantities they sell in are designed for commercial creameries rather than for you.

Clares Carlton Limited,
Wells, Somerset.
Tel: 0749 73900.
— Rennet (in ½ litres),* churns, moulds, chemicals and calico cheese-cloth.

R.J.Fullwood and Bland Limited,
Ellesmere, Salop.
Tel: 069 171 2391.

— Rennet, butter colour and junket tablets.

Chr. Hansen's Laboratory Limited,
476 Basingstoke Road, Reading, Berkshire.
Tel: 0734 861056.

— Rennet, annatto and cultures (all in ½ litres).*

* Note that ½ litre of rennet will be enough for 400 gallons of milk.

US GOAT CLUBS

Alpines International, Stephen Considine, Box 211,
    Oxford, PA 19363.
American Dairy Goat Association, P.O.Box 186,
    Spindale, NC 28160.
American Goat Society, 1606 Colorado Street,
    Manhattan, KS 66502.
Canadian Goat Society, Ottawa, Canada.
Central New York Dairy Goat Society, 6212 Lakeshore
    Road, R.D.4, Clay, NY 13041.
National Nubian Club, R.D.1, Box 416, Glen Gardner,
    NJ 08826.
National Saanan Club, R.D.3, Marysville, OH 43040.
National Toggenburg Club, Box 177A, Strotz Road,
    Rt.1, Asbury, NJ 08802.
New York State Dairy Goat Breeders' Association,
    456 W. Sand Lake Road, Troy, NY 12180.

BACKYARD POULTRY BOOK
Andrew Singer

The author of the Backyard Dairy Book brings the same treatment to the subject of domestic poultry keeping. This book will appeal both to the absolute beginner and the experienced backyarder keen to try out new ideas.

Contents

Stage One: Making the big Decision
Chapter One: Why keep chickens?

Stage Two: Planning and Preparation
Chapter Two: A little technical background
Chapter Three: Which system to use?
Chapter Four: Which hens to buy?
Chapter Five: What to feed them?
Chapter Six: Confining and housing them

Stage Three: Your Chicks Arrive
Chapter Seven: Rearing chicks into layers
Chapter Eight: When to re-stock?
Chapter Nine: The harvest—eggs, meat etc.
Chapter Ten: Diseases and problems

Stage Four: Other Poultry
Chapter Eleven: Ducks
Chapter Twleve: Geese
Chapter Thirteen: Turkeys
Chapter Fourteen: Guinea Fowl, Bantams and Pigeons

"Andrew Singer's comprehensive manual deals with all aspects of poultry keeping and is a must for all potential poultry keepers before making a start"—*Undercurrents*

PRACTICAL SOLAR HEATING
Kevin McCartney

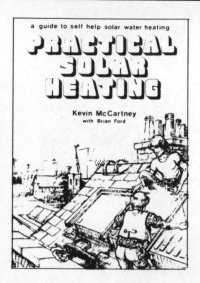

a guide to self help solar water heating

PRACTICAL SOLAR HEATING

Kevin McCartney
with Brian Ford

Even in a climate such as ours up to 50% of our domestic water heating could be supplied by the sun. So far only cost has postponed the widespread use of solar energy in homes throughout the country. Now, however, the rocketing price of conventional energy sources has made solar water heating economically competitive. Furthermore new plumbing materials and techniques now make do-it-yourself installation quite feasible, thereby greatly reducing the capital outlay.

This book has been written by a researcher at the Architectural Association who has designed, built and monitored numerous installations over the past five years. It is the most authoritative, thorough and inexpensive account yet to appear and many of the suggestions and precepts were tested during the successful installation of solar panels at Prism Press.

Contents
1. Solar Energy—what is it, why we should use it and how.
2. Basic Principles—Absorption, heat loss greenhouse effect, heat and temperature, heat capacity and reaction time, tilt and orientation angles.
3. Solar Collectors—Function, types, surface finishes, materials, insulation casing and glazing
4. Storage Tanks—Hot water cylinders, galvanised iron and plastic tanks, heat exchanger cold feed tanks, expansion tanks, pressurised tanks and insulation.
5. Circulation—Thermosyphoning (gravity) and forced (pumped) circulation, pump and pipe sizes and materials, pipe lagging and controls.
6. Mounting the Collectors—Location, building permission, fixing over existing roof removing roof and fixing collectors under glazing bars, wall collectors and free-standing collectors.
7. Choosing a System—Step-by-step guide variations in methods of connecting collector to storage tanks, frost protection and temperature boosters.
8. Plumbing for Solar Systems—Plumbing without a blowtorch, compression fittings, pipe types. With a blowtorch, capillary fittings, low cost plumbing.
9. Swimming Pools—Types of collector required size, location and mounting, insulating the pool effectiveness.
10. Examples—Commercial, Council and D.I. installations.
11. D.I.Y. Collectors—Detailed plans for two types.
12. Installation Guide—Step-by-step instruction
13. Survey of Manufactured Collectors.

SMALL SCALE WATER POWER
Dermot McGuigan

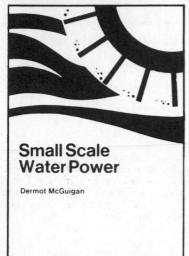

Small Scale Water Power

Dermot McGuigan

Do you like the sound of a rushing brook? It sounds even better when all that energy is harnessed and is lighting and heating your home, or providing the power for a farm or small industry.

If you live near a brook or a river, or are thinking of building near one, this book is a must for you. It tells you how to best tap that source of power to meet your electrical needs.

With the age of cheap energy drawing to a close, interest in water power is growing fast. Those sharing that interest will find this is an excellent sourcebook, telling how to estimate the power in a stream, where the most suitable equipment can be obtained, and at what price.

There are detailed descriptions of working installations, with an analysis of their costs. The recent innovations that have cut the costs of hydropower equipment are explained.

You'll also find here: Information on dams, fish passes, spillways, pipelines, drivers, alternators, governors and legal aspects—plus a detailed manufacturer's index.

SMALL SCALE WIND POWER
Dermot McGuigan

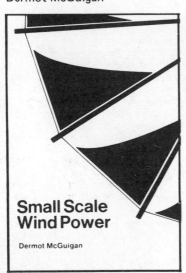

Small Scale Wind Power

Dermot McGuigan

Wind power, harnessed centuries ago by man, is fast gaining new popularity.

Is it for you? This book lets you decide. It presents a survey of numerous working windplants, telling the many purposes which wind power is best suited to serve.

There's an appeal to wind power. It's there— we hear it every day. It's non-polluting and inexhaustible. And it's free, after the initial installation of the windplant.

Homeowners, farmers, small industries are looking again at wind power. Join them. Dermot McGuigan makes it easy to take an informed look. You can reach a decision based on scientific fact, not hopes or promises.

Here, in one book, are all of the pluses and minuses of wind power, and the many types of wind machines now on the market, written so that you can understand them. If you are interested in alternative energy, you'll find this book engrossing. If you want to reduce that growing fuel bill, this book may show you how.

Other Backyard Books available from PRISM PRESS

BACKYARD FARMING
Ann Williams

All you need to know about what you can do with your backyard, whether it is one-third of an acre or ten acres. It covers all purpose layout and fencing, planning your production, equipment that you need, animals to keep and how to feed them off your plot.

BACKYARD SHEEP FARMING
Ann Williams

From breeds and breeding through routine care and problems to using the meat and wool.

BACKYARD PIG FARMING
Ann Williams

Pork and bacon in your freezer. Recycle your waste into edible meat and at the same time let your pig do the digging for you.

BACKYARD RABBIT FARMING
Ann Williams

'Breeding like rabbits' — from four does and one buck you should rear enough rabbit meat for one meal per week in the smallest of areas.

BACKYARD FISH FARMING
Paul Bryant

Turn your aquarium into a food producing tank and watch your food grow in front of you.